HOW TO USE THIS BOOK

FIND OUT ABOUT ME ON PAGE 134!

This LEGO® *Minifigure Handbook* shows you some of the coolest, rarest, and most unusual minifigures ever produced by the LEGO Group. It is divided into 10 themed sections, and each page is a goldmine of minifigure information.

Numbered annotations point out the key features of each minifigure.

This is the name by which the character is most commonly known (many early minifigures did not have official names).

This box is your at-a-glance guide to each minifigure's place in LEGO history.

Quotes from LEGO designers give insights into the minifigure design process.

These boxes feature information about a related minifigure, or a variant of the main featured minifigure.

SAUSAGE SANDWICH BOARD!

ONE OF A KIND

I APPROACH MY JOB WITH RELISH!

❶ Aptly named "sandwich board" piece shaped like a hot dog

❷ Cheerful expression.. Mmm, hot dogs!

❸ Plain tan torso matches bun colour

"The initial design didn't have any mustard, but the final version has just the right amount of zig-zag sauce!"
GITTE THORSEN, LEGO DESIGN MASTER

HOT DOG MAN

This fast-food fanatic was the first collectible Minifigure to dress up as food! The detailed hot dog outfit was brand new in 2015. The sausage, bun, and mustard are all one piece that fits over his minifigure head.

SERVICE WITH A SMILE
Hot Dog Man isn't the only frankfurter fan in the collectible Minifigures theme. The Hot Dog Vendor is also sizzling with excitement about them. He cheerfully serves up a hot dog and soda in Series 17.

MINI STATS

Theme
LEGO® Minifigures
Year
2015
First appearance
LEGO Minifigures Series 13

Rarity

121

1	Not really rare, but still cool!
2	Not too difficult to find, with a bit of searching
3	Appears in only a small number of sets, or in blind polybags
4	Rare – you're lucky if you have this minifigure!
5	Very rare – a legendary minifigure!

This is where you'll find out how rare each minifigure is – or isn't!

LEGO

MINIFIGURE
HANDBOOK

DK

CONTENTS

Welcome to the wonderful world of LEGO® minifigures! You'll find hundreds of colourful characters in this book, on pages packed full of fascinating facts and trivia. They come from all eras, from 1978 to today, so what are you waiting for? Let's dive right in!

LET'S DO THIS!

CHAPTER ONE
EVERYDAY HEROES

HERE'S TO THE DEDICATED MINIFIGURES WHO WORK TIRELESSLY TO HELP OTHERS, EACH AND EVERY DAY!

FIRST TRUE MINIFIGURE

I'M NUMBER ONE!

1 All male minifigures wore hats until male hair pieces appeared in 1979

2 Very simply printed face, with dots for eyes and curved line for a smile

3 Torso decoration is a sticker, not printing

4 Badge identifies him as a member of the LEGO® police force

POLICEMAN

MINI STATS

Theme
LEGOLAND® Town

Years
1978, 1981

First appearance
Police Car (600)

Rarity

With posable arms and legs, and a printed face, this policeman was the first true minifigure. Before him, LEGO characters were mostly static figures with little playability. He is too big to fit into his police car, though, so he has to sit on its bonnet!

BIRTHDAY REBOOT

To celebrate his 40th birthday in 2018, the 1978 Police Officer got a fresh look and his own mini version of the classic LEGO set he first appeared in! He has a new, printed torso in Minifigures Series 18, but the same happy smile.

MINI STATS

Theme
LEGO® City

Year
2015

First appearance
Service Truck (60073)

Rarity

Where would LEGO City be without its construction workers? In a big mess, that's where! This minifigure joined the army of orange-suited workmen in 2015. Perhaps due to updated health and safety regulations, he is the first to wear a helmet with built-in ear defenders!

UPDATED CLASSIC

CONSTRUCTION WORKER

1 Many LEGO construction workers now wear this helmet

2 Face printed with stubble and a raised eyebrow

3 Torso design shows a hooded top under work clothes

4 Grey hands represent gloves

FILM STAR

THE LEGO® MOVIE™ introduced us to the most famous construction worker of all: the lovable Emmet!

A PROUD TRADITION

LEGO construction workers date back to the earliest minifigure days. This well-dressed worker was one of the first, appearing in 1979's Street Crew (set 542).

7

The Diner Waitress has a real retro look, with her horn-rimmed specs and 1950s-style hair piece. What really makes her stand out, however, are her accessories – never before had a minifigure had such a cool set of wheels but no car!

HOW MAY I HELP YOU?

Diner Waitress is the only Series 11 Minifigure with a double-sided head piece. One side of her face has a wide smile, while the other has an annoyed frown.

MINI STATS

Theme
LEGO® Minifigures

Year
2013

First appearance
LEGO Minifigures Series 11

Rarity

DINER WAITRESS

NAMESAKE
The Diner Waitress is named "Tara", after Tara Wike: the LEGO designer she is based on!

① Soft-serve ice-cream-style hairdo

② Double-sided head has two expressions

③ Ice-cream sundae balanced on a white serving tray

④ Rockin' pink roller skates!

LEGO DESIGNER'S SECOND JOB!

FIRST PRINTED TORSO

TICKETS, PLEASE!

TRAIN CONDUCTOR

① Red cap first seen on 1975's non-posable LEGO figures

② Blue torso with jacket and tie printing featured in more than 30 sets up to 2003

③ Classic posable leg design has remained unchanged since 1978

All aboard!
The train conductor was the first minifigure to have a printed torso. Previously, torsos had been stickered or left plain, but now designs were printed directly onto the torso piece.

AN ENDURING CHARACTER

Two years after the Train Conductor first appeared, a new Train Guard arrived on the scene. He wore the same hat, but had a different torso design and blue trousers. He was included in more than a dozen sets over the course of the next six years.

MINI STATS

Theme
LEGOLAND® Town

Year
1978

First appearance
Car Transport Wagon (167)

Rarity

I'M CREATING A NEW ACRYLONITRILE BUTADIENE STYRENE MOLECULE!

❶ Hairstyle piece has a ponytail at the back

❷ Reversible head has a serious look on one side and a shocked expression on the other

❸ Torso printing features pens and an ID card

❹ Lab coat print continues on back of the torso

CHEMIST

MINI STATS

Theme
LEGO® Ideas

Year
2014

First appearance
Research Institute (21110)

Rarity

Take a closer look at the world with one of the three exclusive scientists included in the Research Institute set. This was chosen to become a real LEGO set from hundreds of fan-built models submitted to the LEGO Ideas website.

REAL SCIENCE

The Chemist is based on Dr Ellen Kooijman, the real-life geoscientist who came up with the idea for the Research Institute set.

LAB PARTNERS

The Chemist is joined in the Research Institute by a Paleontologist studying a huge dinosaur skeleton, and an Astronomer with a telescope.

MINI STATS

Theme
LEGO® Minifigures

Year
2011

First appearance
LEGO Minifigures
Series 4

Rarity

Hazmat Guy's job is to deal with hazardous materials, so it's understandable that he looks so anxious! Luckily, he is protected by his airtight hazmat helmet, with its built-in visor, breathing apparatus, and radiation warning symbol.

DANGEROUS WORK
Hazmat Guy is not the only one with a dangerous job. The Dino Tracker from Minifigures Series 13 seems to enjoy hers, though!

HAZMAT GUY

HOW WILL I EAT MY LUNCH IN THIS SUIT?

1 Hazmat helmet with wide, clear visor

2 Anxious face with printed stress lines!

3 Hazard icon printed on helmet bib, and on torso beneath

4 Spray gun with hose fits onto back of helmet

CAMEO
Some Robo SWATS in THE LEGO® MOVIE™ wear hazmat suits, too.

HI-TECH HELMET

STATS

Theme
LEGOLAND® Town

Year
1978

First appearance
Ambulance (606)

Rarity

The first-ever female minifigure works as a nurse aboard the LEGOLAND Town Ambulance. Released in 1978, she is also the only medic minifigure to have a sticker on her torso instead of printed details.

FIRST FEMALE

NO ROOM
The nurse's ambulance is too small for her to fit inside, but she can ride on the bonnet or the roof.

NURSE

AT LAST! WITH LEGS, I CAN FINALLY SIT DOWN!

1 Same hair piece worn by non-posable LEGO figures from 1975

2 Head piece has the original simple face printing on it

3 Plain white torso with red cross sticker

4 Posable legs – still a novelty in 1978!

WELL EQUIPPED

In 1980, a variant of this medical minifigure appeared in Paramedic Unit (set 6364). Her torso is printed with details including a stethoscope and a pocket with a pen.

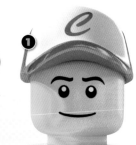

COVERS ALL THE BASES!

TAKE ME OUT TO THE BALL GAME!

BASEBALL PLAYER

1 Curved-bill hat also worn by the Rapper, Lumberjack, and Baseball Fielder Minifigures

2 Baseball bat also carried by Harley Quinn from LEGO® DC Comics Super Heroes

3 "Clutchers" team name refers to the "clutch power" that holds LEGO bricks together

4 Plain white legs are unique, thanks to belt printing on hips

This sports hero
was the first to wield a
LEGO baseball bat, but he was
not the first LEGO baseball player. In 1999, a special
Boston Red Sox minifigure was given away at the
team's Fenway Park stadium. He also wears a red cap
and a white uniform, and is now very rare indeed!

PERFECT CATCH

In 2013, a Baseball Fielder with a "Stackers" uniform and a catcher's mitt instead of a left hand featured as part of Minifigures Series 10.

MINI STATS

Theme
LEGO® Minifigures

Year
2011

First appearance
LEGO Minifigures Series 3

Rarity

NOW, WHAT CAN I COOK UP FOR YOU TODAY?

1 Traditional chef's hat is still in use today

2 All minifigures had the same smiling face until 1989

3 Most early torso prints used just one or two colours

4 This shade of grey no longer exists in the LEGO® colour palette

MINI STATS

Theme
LEGOLAND® Town

Years
1979–1980, 1982–1983, 1985, 1991

First appearance
Snack Bar (675)

Rarity

CHEF

FIRST CHEF'S HAT!

PIZZA SPECIAL

Behind door 21 of the 2005 LEGO® City Advent Calendar (set 7324) is the Pizza Chef minifigure. Wearing the iconic chef's hat and whites, he's ready to serve some tasty pizza. Judging by the wide smile under his curly moustache, he likes his job!

In 1979, LEGOLAND Town got its very first snack bar. Now hungry minifigures could fill their bellies with the many delicious treats cooked by the hardworking Chef. This Chef's special offering is his hat – he was the first minifigure to wear one like it!

MINI STATS

Theme
LEGO® City

Year
2020

First appearance
Racing Cars
(60256)

Rarity

This speedster zoomed into LEGO City in 2020 powered by Octan Energy. He is ready for some fuel-filled fast-lane action behind the wheel of racecar 29 – a number that also appears on the back of his racing jacket.

SUPER-SPEEDY RACER

OCTAN RACECAR DRIVER

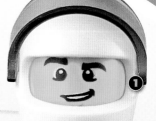

HAVE YOU GOT THE ENERGY TO RACE ME?

1 Face print used for four fellow racers in the LEGO® Speed Champions theme

2 Racing jacket with white collar securely fastened

3 New torso print features an Octan Energy symbol

4 Grey hands look like racing gloves

VROOM FOR TWO?

Race Car Guy from Minifigures Series 18 takes the term "motorhead" to a whole new level. He loves his red racecar so much, he wears it everywhere he goes!

Theme
LEGOLAND® Town

Years
1982–1983, 1985, 1991

First appearance
Post Office (6362)

Rarity

LEGOLAND Town got its first delivery service in 1982 with a Post Office (set 6362) and Post Office Van (set 6651). Both sets feature this jolly Postman, who has a red torso printed with the bugle-shaped post office logo on the left-hand side, and matching black hat and legs.

FIRST **MAIL** MINIFIG

POSTMAN

I'M SURE TO GET LOTS OF FAN MAIL!

1 Another postal worker and a Fire Captain also wore this new black hat in 1982

2 The LEGO post-office logo is a bugle or post horn

3 Black is the most common colour for minifigure legs, appearing in more than 1,000 sets!

MODERN MAIL

In 2008, the LEGO City postal service zoomed into the modern world with Air Mail (set 7732), part of the Cargo subtheme. The set includes an Air Mail Worker wearing cool sunglasses and a cap.

FULLY BOOKED!

> I'VE ALWAYS WANTED TO BE IN A BOOK!

1 Hair mould first seen on Ginny Weasley from LEGO® Harry Potter™

2 Mug has a message for noisy visitors to the library

3 Book accessory can be opened and closed

TITULAR TITTERS

The book title *Oranges and Peaches* is a joke mishearing of *On the Origin of Species* by Charles Darwin.

Oranges and Peaches

LIBRARIAN

This bespectacled bookworm features all-unique printing, including the "Shhh!" on her coffee mug! Part of the range of collectible Minifigures launched in 2013, she has her own online bio that describes her love for books of all kinds – including ones about minifigures, no doubt!

BOOK LOVER

Here's another minifigure who turned her passion for books into her career! The curly-haired Bookseller works in the cosy LEGO Creator Bookshop (set 20170).

MINI STATS

Theme
LEGO® Minifigures

Year
2013

First appearance
LEGO Minifigures Series 10

Rarity

HELPING OTHERS IS MY DEPARTMENT.

FLAME-HAIRED FIRE CHIEF

1 Wavy ponytail hair shared with Barbara Gordon in THE LEGO® BATMAN MOVIE sets

2 New freckled face print with green glasses

3 LEGO City fire-department jacket features a gold flame logo

1 MUGSHOT
Freya has a special mug that features a flaming toasted marshmallow head!

FREYA McCLOUD

When there's a blaze in LEGO City, the first person on the scene will be Freya McCloud – the fast-thinking Fire Chief. The star of the LEGO *City Adventures* TV series first appeared as a minifigure in her roaring Fire Chief Response Truck.

MINI STATS

Theme
LEGO® City

Year
2019

First appearance
Fire Chief Response Truck (60231)

Rarity

TEAM EFFORT
In heavy-duty overalls, these firefighters team up to battle a blaze in Burger Bar Fire Rescue (set 60231). Let's hope they get some flame-grilled burgers as a reward!

MINI STATS

Theme
LEGO® Minifigures

Year
2012

First appearance
LEGO Minifigures
Series 6

Rarity

The Surgeon is dressed to operate in her surgical scrubs. She was the first minifigure to wear her scrubs cap piece, and is unusual because you cannot see her mouth beneath her printed mask. However, her friendly eyes suggest there is a traditional LEGO smile underneath it, ready to reassure her patients.

MEDIC IN A MASK

SURGEON

THIS WON'T HURT A BIT!

1 Scrubs cap also worn as a shower cap by the Shower Guy Minifigure

2 White hands look like surgical gloves

3 Syringe piece also used by Nurse from Minifigures Series 1

4 X-ray piece has the same proportions as a LEGO Skeleton figure

SCREEN SURGEON

A similar surgeon is among the Master Builders in THE LEGO® MOVIE™.

AMBULANCE DRIVER

Look out for this brave paramedic when the LEGO City ambulance whizzes by. She safely delivers emergency patients to doctors at the LEGO City Hospital (set 60204).

There'll be no slipping on banana peels in LEGO City when this unsung hero is around! The Street Sweeper keeps the brick streets clean and tidy in a special sweeping vehicle, wearing protective construction overalls, a blue cap, and grey gloves.

THIS ONE'S A SWEEPER!

MINI STATS

Theme
LEGO® City

Year
2020

First appearance
Street Sweeper (60249)

Rarity

STREET SWEEPER

IF I SEE IT, I SWEEP IT!

1 Curved-bill cap protects his eyes from sun and dust

2 Face print first seen in 2019 on a LEGO City Space Port astronaut

3 Grey hoodie for added warmth under orange overall

4 Utility belt holds a walkie-talkie on other side of torso

VITAL ROLE

The first LEGO street sweeper appeared in a 1978 LEGOLAND Town set.

DUMPSTER DUO

The Street Sweeper works solo, but this terrific team is never far behind him, collecting trash in their Garbage Truck (set 60220).

SHE DIGS BONES!

PALEONTOLOGIST

THIS ISN'T A NEW AMMONITE. IT'S 65 MILLION YEARS OLD!

1 Helmet and hair are one piece

2 Two-tone printed arms look like short sleeves

3 New ammonite fossil accessory

4 Legs printed with shorts, white socks, and boots

"My favourite part of this minifigure is the helmet designed by my colleague Gitte Thorsen – it's awesome!"
CHRIS B. JOHANSEN, LEGO DESIGN MASTER

Dressed in safari gear, the Paleontologist is on a mission to dig up some dinosaurs! She's made a good start – she was the first minifigure to find her fossil piece. Let's hope she shares her discovery with the original LEGO paleontologist from 2014's Research Institute set!

MINI STATS

Theme
LEGO® Minifigures

Year
2015

First appearance
LEGO Minifigures Series 13

Rarity

BONE HISTORY

The Paleontologist's bone first appeared in 2011 and since then has featured in more than 100 different sets, including the Cave Woman from Minifigures Series 5.

WANT A HAND?

HEROIC HANDYMAN

HARL HUBBS

1. Tousled hair in a colour first seen on soccer player Mario Götze's minifigure

2. Leaking pen in shirt pocket!

3. Leather belt print – he wears a detachable belt here in the Garage Centre (set 60232)

4. Hand-stitched patch on well-worn dungarees

MINI STATS

Theme
LEGO® City

Years
2019–2020

First appearance
Garage Centre (60232)

Rarity

No job is too big or small for helpful handyman Harl Hubbs. His patched-up, paint-streaked dungarees are testament to how hard he works around LEGO City. His dual-sided head has a frown too, for particularly tricky jobs.

SLICK SIDEKICK

Tread Octane appears alongside Harl Hubbs in the LEGO City Tuning Workshop (set 60258), sporting equally high hair and helpful hands. Like Harl, he is skilfully servicing his motorcycle there – a suitably cool hot rod.

22

MINI STATS

Theme
LEGO® Minifigures

Year
2019

First appearance
LEGO Minifigures
Series 19

Rarity

By the look on the Dog Sitter's face, one of her dogs has just had a call of nature! Looking after dogs for a living isn't all cuddles and playing fetch. This Minifigure is dressed for dirty work in a striped T-shirt, dungarees, and a special paw-print logo cap that pins her long hair up in a ponytail.

POOPER-SCOOPER

DOG SITTER

DO I SMELL POO?

EW!
The Dog Sitter comes with a freshly swirled poo. The new 1x1 piece later appeared in white as a shell and an ice cream.

① Paw-print baseball cap and ponytail are one piece

② Two-tone printed arms look like a short-sleeved T-shirt

③ Leg printing features short denim dungarees and dangerously white sneakers!

④ Shovel seen in bright green for the first time

SMILING SITTER

The first LEGO sitter appeared in Minifigures Series 16. The Babysitter comes with a babbling baby minifigure, who looks equally happy to see a bottle of milk!

23

This hardworking **Minifigure's** work is never done. He is up early at sunrise, day in and day out. The farmer can be found cleaning out pig pens, milking cows, and feeding hens in his dungarees and check shirt, which are as hardwearing as he is!

SPOT THE NEW PIG

MINI STATS

Theme
LEGO® Minifigures

Year
2016

First appearance
LEGO Minifigures
Series 15

Rarity

FARMER

> MY JOKES ARE CORNY!

CROP COSTUME

If Corn Cob Guy isn't careful, the Farmer will harvest him along with the rest of his crops! This corn enthusiast's yellow and green kernel costume has a sandwich-board head piece that hangs past his hips.

1 Wide-brimmed hat seen in brown for the first time

2 Green check shirt under dungarees

3 Spotted pig companion, first seen with the Farmer

4 Heavy-duty brown boots printed on legs

BRILLIANT BRAIN

PIZZA WAS THE GREATEST INVENTION EVER!

J.B.

1 Short lavender hair with built-in scientific goggles

2 Beaming face – she loves her job, and pizza!

3 Frankenstein motif T-shirt under white lab coat

SPACE SCIENTIST

While J.B. has a passion for the paranormal, this LEGO City scientist studies extraterrestrial specimens in a Space Port Minifigure blister pack (set 40345). Her safety goggles are printed on her face.

MINI STATS

Theme
LEGO® Hidden Side

Year
2019

First appearance
J.B.'s Ghost Lab
(70418)

Rarity

This enthusiastic scientist eats, sleeps, and works in her Ghost Lab. She made her first LEGO set appearance there, dressed in her lab coat – and with her dinner in hand! One of a team of ghost hunters in LEGO Hidden Side sets, J.B. is the brains behind the team's technology.

25

MY JOB STINKS, BUT I LOVE IT!

FIRST LEGO SKUNK

ANIMAL ID
The Animal Control badge on her shirt features a raccoon.

1 White mesh net first seen with the Animal Control Minifigure

2 Green utility shirt with name tag and identification badge

3 Specially designed khaki trousers with brown side stripe

ANIMAL CONTROL

Plagued by pesky LEGO critters? Call Animal Control! This specialist Minifigure is trained to collect animals in her enormous net and return them to the wild. She is clearly good at her job – she caught the very first LEGO skunk!

EQUIPPED FOR THE JOB

Another Minifigure with a not-so-glamorous job appeared in Series 16 with a new accessory. The Janitor has a realistic-looking mop head attached to a bar piece.

Theme
LEGO® City
Year
2019
First appearance
Donut Shop Opening
(60233)

Rarity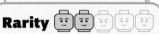

The citizens of LEGO City love good coffee, and this minifigure is there to serve it to them! Dressed in an espresso-coloured apron emblazoned with a happily steaming LEGO coffee cup, this Barista delivers his coffee on a handy cargo bike. He is part of a bevy of expert baristas who arrived in LEGO City from 2015.

COFFEE LOVER

BARISTA

I TAKE MY COFFEE VERY SERIOUSLY.

1 Side-swept hair shared with the Skater from Minifigures Series 1

2 Buttoned-up barista shirt with a mocha collar

3 Name tag attached to an apron strap

4 Apron strings are tied at the back of his torso

FRESHLY GROUND
The first LEGO barista minifigure was Larry the Barista from THE LEGO® MOVIE™.

IN DEMAND
A second barista serves coffee, croissants, and cake to the masses on the City Coffee cargo bike in the Donut Shop Opening set. The Female Barista wears the same uniform.

Need a ticket for the Ghost Train Express? See Ms Santos— she's the ticket seller in the set. She is dressed to greet travellers in her train company uniform with a neck scarf, name tag, and ID badge. Unfortunately, Ms Santos is also forced to welcome a ghost – aboard her body! It possesses her, transforming her minifigure into a green-haired Gloombie.

GHOST-TRAIN GLOOMBIE

MINI STATS

Theme
LEGO® Hidden Side

Year
2019

First appearance
Ghost Train Express
(70424)

Rarity

MS SANTOS

WELCOME ABOARD!

1 Long, wavy hair piece tumbles over her torso

2 Striped waistcoat and crisp white shirt under suit

3 Burgundy suit trousers

POSSESSED
This Gloombie head and wild green hair can be switched for Ms Santos's regular hair and head when she is possessed by a ghost.

TEACHER TROUBLE
No one is safe from the LEGO Hidden Side ghosts. Mr Clarke is a teacher in the Newbury Haunted High School (set 70425) who also becomes a Gloombie!

CHAPTER TWO
YOU'RE HISTORY.

TRAVEL BACK IN TIME TO SEE HOW MINIFIGURES HAVE PLAYED THEIR PART ALL THROUGH THE HISTORY OF THE WORLD!

MINIFIGURES, PREPARE FOR GLORY!

1 Bronze helmet colour first seen on this minifigure

2 Torso print is a cuirass – a muscly piece of armour

3 Spear also carried by the Tribal Chief from Minifigures Series 3

4 Crimson fabric cape

SPARTAN

Theme
LEGO® Minifigures

Year
2010

First appearance
LEGO Minifigures
Series 2

Rarity

A brave and fearsome warrior, the Spartan features all-unique printing, right down to his sandals! He carries a new, rubber-tipped spear and a shield that can be decorated with other LEGO pieces.

GREEK AND UNIQUE!

HISTORICAL HEADGEAR

The Spartan's distinctive helmet is the same as the one seen on the Temple Statue in the LEGO® Atlantis theme, only in bronze.

30

 **MINI STATS**

Theme
LEGO® Pirates

Years
1989, 1991, 1993, 1995–1997, 2001–2002

First appearance
Forbidden Island (6270)

Rarity

In 1989, LEGO Pirates introduced the very first minifigures with different faces and body parts. Captain Redbeard has a peg leg, a hook for a hand, a printed eyepatch, and a beard. After him, minifigures were never the same!

FIRST OF A NEW BREED!

CAPTAIN REDBEARD

WALK THE PLANK!

1 Only this variant of Redbeard has printing on his hat

2 No minifigure had ever had facial hair or an eyepatch before!

3 New epaulet element over printed torso

4 Hook functions just as well as a regular minifigure hand

① ② ③ ④

COMIC CAPERS

A LEGO Pirates comic book, *The Golden Medallion*, was released in 1989.

CAPTAIN BRICKBEARD

In 2009, a new captain commanded the pirates. Brickbeard also has an eyepatch, hook-hand, and peg leg. Could he be Captain Redbeard reborn?

The glitzy **Gold Knight** is a hero from the Fantasy Era of the LEGO Castle theme. He must have used magic to strengthen his armour and weapons, as gold is normally a soft metal – good for decoration but not for fighting!

MINI **STATS**

Theme
LEGO® Castle

Year
2009

First appearance
Drawbridge Defense (7079)

Rarity

KNIGHT SHINING BRIGHT!

GOLD KNIGHT

I'M FROM THE GOLDEN AGE!

SKELETON CREW

Drawbridge Defense also includes three spooky foes for the Gold Knight to battle: a white skeleton, a black skeleton, and a scythe-wielding skeleton riding – you guessed it – a skeleton horse!

① Pointed visor and helmet are separate pieces

② Gold chrome sword

③ Unique breastplate

WILD WEST HERO

SHERIFF

> STICK 'EM UP, STRANGER!

1 Hat mould first used in LEGO® *The Lone Ranger*™ sets

2 Moustache fits between head and torso

3 Gold chain for a pocket watch crosses brown vest

4 Leather holster print on right leg

'ELLO 'ELLO 'ELLO

A fellow moustachioed enforcer of the law appeared in Series 11 of the Minifigures theme! Dressed all in black with a unique police helmet, the Constable walks the brick beat with a traditional truncheon.

MINI STATS

Theme
LEGO® Minifigures

Year
2015

First appearance
LEGO® Minifigures
Series 13

Rarity

There are many things about the Sheriff that command respect: the Stetson hat, the star badge, the moustache... it's no surprise that he is one of the most revered characters in the Minifigures theme. Just make sure you're not on his Most Wanted list!

HISS!

SHE RULES IN JEWELS!

1 Smooth, printed hair piece

2 Torso print shows elaborate hawk and winged snake jewellery

3 Green snake seen in several LEGO® NINJAGO® sets

4 White slope piece with ornate printed detail

ROYAL RELATION

The appearance of the Egyptian Queen is inspired by that of Queen Cleopatra VII, the last pharaoh of Egypt.

MINI STATS

Theme
LEGO® Minifigures

Year
2011

First appearance
LEGO Minifigures Series 5

Rarity

This regal Minifigure boasts a special hair piece printed with a winged scarab. The rest of her printing is also unique to her. To give the impression of a dress, she has a sloped piece on her lower half in place of standard minifigure legs.

A PAIR OF PHARAOHS

The first ancient Egypt-inspired character in the Minifigures theme was the Pharaoh from Series 2. Could he rule alongside the Egyptian Queen?

MINI  STATS

Theme
LEGO® City

Year
2018

First appearance
Capital City (60200)

Rarity

Visit the LEGO City museum and you will see the Museum Caveman exhibit. Stay there long enough and this beardy man from the prehistoric past may even grunt "hello"! He could look familiar, as he shares his head with LEGO NINJAGO hero Cole. Or is he a Ninja in disguise?

INTERACTIVE **EXHIBIT**

CAVEMAN

1 Overgrown hair and beard are one element, first seen in 2010

2 Bushy eyebrows and fixed scowl beneath the face fur!

3 Clothing made from animal hides and bone

4 Tan legs and reddish-brown hips combination first worn by this minifigure

FUZZY FIRSTS

The club-wielding Caveman from Minifigures Series 1 was the first prehistoric man to enter the modern world of LEGO sets. He was also the first minifigure to have a unibrow!

The **Forestman began** life in 1987 as the leader of six outlaws guarding a treasure chest in a forest hideout. His companions wear blue, black, or red feathers in their matching caps – a design first seen on them. All six look like merry men!

FIRST
TO WEAR
GREEN

MINI STATS

Theme
LEGOLAND® Castle
Years
1987–1988
First appearance
Camouflaged Outpost (6066)

Rarity

FORESTMAN

1 Feathers of different colours and sizes can be added to this hat design

2 Quiver of arrows fits between head and torso

3 The Forestmen were the first minifigures to wear green

4 Money pouch on waistband

YOU MAY BE OLDER SIR, BUT WHERE IS YOUR MOUSTACHE?

REDESIGNED ROGUE

A more detailed Forestman with a goatee beard, thin moustache, and cheeky smile was included in the first series of LEGO Minifigures in 2010.

GOLDEN FACE MASK

UNMASK ME, I DARE YOU!

MUMMY QUEEN

1 Turquoise headdress with gold and red trim

2 Golden painted face with blue eyes and red lips

3 Ornate scarab necklace collar continues on back of torso

4 Leg wrappings and gold-trimmed skirt

YIKES!
Peek beneath the golden face mask (or turn the dual-sided head) to reveal the Mummy Queen's corpse face!

The shimmering, serene Mummy Queen is an ancient sight to behold. Unearthed in Minifigures Series 19, she is beautifully detailed, with a golden face mask and jewellery, and printing on the sides of her arms and legs.

EGYPTIAN HERITAGE

There are several Egyptian Minifigures in the Minifigures theme. The Egyptian Warrior from Series 13 brandishes a geometric-patterned shield. Did he once serve the Mummy Queen?

MINI STATS

Theme
LEGO® Minifigures

Years
2019

First appearance
LEGO Minifigures Series 19

Rarity

37

KNIGHT KNIGHT!

1 Visor clips onto helmet and can swivel up and down

2 Helmet is shared with the classic red spaceman

3 Printed shield emblem

4 Black legs with red hips seen on more than 50 minifigures

FOUR KNIGHTS
This knight's torso design is used on just four other minifigures.

This brave knight is one of a matching pair released with two knights from a rival faction, who wear different helmets, colours, and torso designs. The set also includes a sword, an axe, a hatchet, and a shield.

FRIGHT KNIGHTS
The terrifying Fright Knights appeared in 1997, ruled not by a king but by the dreaded Basil the Bat Lord – who rides a dragon called Draco!

MINI STATS

Theme
LEGO® Castle

Year
1983

First appearance
Castle Figures (6002)

Rarity

MINI STATS

Theme
LEGO® Vikings

Year
2006

First appearance
Vikings Chess Set
(851861)

Rarity

The LEGO Vikings Chess Set was released in 2006, alongside two other Vikings sets. To avoid getting the minifigures mixed up, which would make it hard to use them as chess pieces, some of their elements were glued together. In the set, the Red King and his army go into battle against a matching team in blue.

VIKING CHESS PIECE!

RED KING

THIS IS NO GAME!

1 Horns are glued to helmet

2 Red fabric cape also worn by Red Queen

3 He wields a golden sword on the chessboard

MATCHING
The Viking chess pieces have the same torso design as the dwarves in 2008 Castle sets.

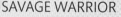

SAVAGE WARRIOR
The Barbarian Armour Viking was one of the first warriors released in the LEGO Vikings line in 2005. With ragged armour and a raging expression, he is best avoided on the battlefield!

YOU'RE HISTORY!

39

Although famously part of a trio, this LEGO musketeer is one of a kind. A legendary fighter, he appeared solo in LEGO Minifigures Series 4, proudly brandishing his silver rapier. The Musketeer has a goateed face and special leg printing that depicts long boots. He shares his hat mould with the equally dashing Captain Barbossa in the LEGO® *Pirates of the Caribbean*™ theme.

Theme
LEGO® Minifigures

Year
2011

First appearance
LEGO Minifigures Series 4

Rarity

ALL FOR BRICKS
AND BRICKS
FOR ALL!

MUSKETEER

1 Jaunty hat mould was new in 2011

2 Rapier piece also functions as an antenna in some LEGO sets

3 Historical French fleur-de-lis symbol

4 Musketeer wears gloves for protection

HEROIC HATS ON

ON GUARD!

The Musketeer met his match in 2014. The Swashbuckler from LEGO Minifigures Series 12 features the same rapier and hat piece, in different colours.

KENDO FIGHTER

MY SWORDS GIVE YOU SPLINTERS!

1 Kendo mask mould, first seen in LEGO® NINJAGO® sets, has a silver grille

2 Wooden weapons are katana swords in tan for the first time

3 Grey stomach and chest protector printed on torso

SKILLED SWORDSMAN

His twin swords
may be made from bamboo, but in the minifigure hands of this master of martial arts, they are formidable weapons. The Kendo Fighter is dressed in traditional training armour, with a lightweight chest plate and protective flaps on his legs.

ON THE OFFENSIVE

The Kendo Fighter wasn't the first masked swordsman in the Minifigures theme. The Fencer thrust his fencing foil in Series 13, dressed in a padded suit and safety mask.

MINI STATS

Theme
LEGO® Minifigures

Year
2016

First appearance
LEGO Minifigures Series 15

Rarity

41

I AM VERY AMUSED!

NEW SKIRT!

PRINTS CHARMING

The Queen's skirt printing is inspired by playing-card patterns, with both hearts and diamonds.

1 Kindly expression first seen on this minifigure

2 Short printed cape resembles traditional royal fur

3 Unique skirt piece continues torso printing

QUEEN

Theme
LEGO® Minifigures

Year
2016

First appearance
LEGO Minifigures
Series 15

Rarity

The first minifigure to wear a sloped skirt was a LEGO Castle Maiden in 1990, but the Queen of LEGO Minifigures requires a more elaborate look. This new piece is 2x4 studs in size, so hopefully the doors of the Queen's castle are equally as wide!

CAPED COUPLE

The Queen rules with the Classic King from LEGO Minifigures Series 13. They both wear two capes, neither of which is seen on any of their minifigure subjects.

MINI

MINI STATS

Theme
LEGO® Adventurers

Years
1998–1999

First appearance
Adventurers Tomb
(2996)

Rarity

LEGO Adventurers was a new theme for 1998, centred on the search for the Re-Gou Ruby in Egypt. Pharaoh Hotep was the creepy king who guarded the ruby, and the first-ever LEGO mummy! He has unique head, leg, and torso printing.

ARE YOU MY MUMMY?

OH MUMMY!
The word "mummy" comes from the ancient Persian word for an embalmed body. Yuck!

PHARAOH HOTEP

 Ancient Egyptian headdresses such as these are known as "nemes"

 Rare minifigure nostrils!

 Blue segmented body armour

4 Toe print on leg piece

THAT WRAPS IT UP!

The LEGO® Studios theme got in on the mummy action in 2002 with Curse of the Pharaoh (set 1383). Its mummy has a bandaged head, torso, and legs, and a less colourful take on Pharaoh Hotep's headdress.

43

This dashing knight can only be found in the Black Knight's Castle set, along with three other knights wearing plumes of red, blue, and yellow respectively. All four knights have their own lance, sword, and tapering kite-shaped shield, decorated with a multicoloured dragon design.

 *BEAUTIFUL PLUMAGE!*

MINI **STATS**

Theme
LEGO® Castle

Year
1992

First appearance
Black Knight's Castle
(6086)

Rarity

WHITE PLUME KNIGHT

JOUST A MINUTE!

① Elegant dragon-shaped plume

② Plumes on both sides of helmet

③ Pointed visor moves up and down

④ Breastplate armour piece over blue torso with printed armour

ANGRY EYES

This mean-looking Evil Knight with a kite-shaped shield was included as part of LEGO Minifigures Series 7 in 2012. The boar on his shield mirrors his angry expression!

44

ANOTHER KIND OF NINJA!

FEMALE NINJA

> **I'M GREEN AND RARELY SEEN!**

1 Head printed with long eyelashes and a headband

2 A clip on the back of the head wrap can hold a katana sword

3 Shuriken and dagger hidden in printed robe

MIRROR NINJA

The green Male Ninja wears his dagger and shuriken (throwing star) the other way around from the Female Ninja.

One of the first

female characters to appear in the LEGO Castle Ninja subtheme, this well-armed warrior comes in a three-minifigure set with a green Male Ninja and a Samurai Lord. Each has a display base that also holds a collector's card.

EVERYTHING'S GONE GREEN

In LEGO® NINJAGO®, Lloyd Garmadon becomes the legendary Green Ninja, destined to win the great battle between good and evil – if he can master the art of Spinjitzu first!

MINI STATS

Theme
LEGO® Castle

Year
2000

First appearance
Mini Heroes Collection Ninja #3 (3346)

Rarity

45

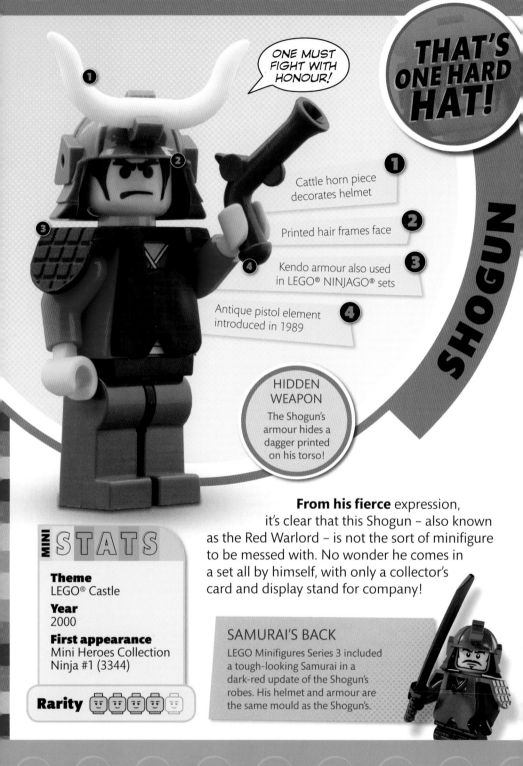

ONE MUST FIGHT WITH HONOUR!

SHOGUN

1 Cattle horn piece decorates helmet

2 Printed hair frames face

3 Kendo armour also used in LEGO® NINJAGO® sets

4 Antique pistol element introduced in 1989

HIDDEN WEAPON
The Shogun's armour hides a dagger printed on his torso!

From his fierce expression, it's clear that this Shogun – also known as the Red Warlord – is not the sort of minifigure to be messed with. No wonder he comes in a set all by himself, with only a collector's card and display stand for company!

MINI STATS

Theme
LEGO® Castle

Year
2000

First appearance
Mini Heroes Collection
Ninja #1 (3344)

Rarity

SAMURAI'S BACK
LEGO Minifigures Series 3 included a tough-looking Samurai in a dark-red update of the Shogun's robes. His helmet and armour are the same mould as the Shogun's.

In minimal armour and wearing only a loincloth, the Roman Gladiator can move with speed and agility in battle against an opponent. He waits for the perfect moment to strike with his gold trident. He is the second Gladiator in the Minifigures theme – the first appeared in Series 5, in a heavy gold helmet.

LIGHT-WEIGHT WARRIOR

ROMAN GLADIATOR

THIS DOUBLES UP AS A FORK.

1 Coiled hair in dark brown for the first time

2 New face with stubble and a serious expression

3 Muscular torso with brown leather shoulder guard

4 Loincloth printed on hips and legs

LION ARM

The Gladiator has a bronze lion on his right arm. He is nicknamed "The Lion" because of his courage and strength.

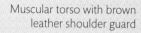

ROMAN RULER

The Roman Emperor from Minifigures Series 9 has gold laurel leaves in his hair and clutches a scroll written in Latin.

VENI, VIDI, VICI

MINI STATS

Theme
LEGO® Minifigures

Years
2015

First appearance
LEGO Minifigures
Series 13

Rarity

Armed with a pair of katana swords, the Samurai has spent many years training to master them. Unlike previous Samurai minifigures, this warrior appears without a helmet. Her fearsome expression – and even more fearsome skill – means her foes know to stay far away!

FEARED FIGHTERS

The Samurai wasn't the first female fighter in the LEGO Minifigures theme. She joined the ranks of Forest Maiden, Battle Goddess, and Viking Woman.

SAMURAI

TWO SWORDS ARE BETTER THAN ONE.

OUT OF ACTION

The Samurai wears a simple robe with a sash under her armour, in bright red.

1 Smooth bun first seen in 2011 on the Sumo Wrestler

2 Armour appears in dark red and is printed for the first time

3 Printing extends the armour design onto legs

NEW PRINTED ARMOUR

48

CHAPTER THREE
OUT OF THIS WORLD

MINIFIGURES HAVE BEEN EXPLORING SPACE SINCE 1978. CHECK OUT THIS COSMIC COLLECTION OF CLASSICS!

MINI STATS

Theme
LEGO® Space

Years
1993–1994

First appearance
Celestial Sled (6834)

Rarity

Befitting his status as leader of a team of hot-shot civilian scientists, this moustachioed minifigure has a unique torso with a formal jacket print. His orange visor was new for the LEGO Space Ice Planet 2002 subtheme and protects the Chief's distinctive white whiskers from the cold.

HAIR WEAR

The Chief isn't the only Ice Planet character with distinctive hair. Ice Planet Woman has fiery red locks, while Ice Planet Man sports a blond fringe.

ICE PLANET CHIEF

> A FROZEN MOUSTACHE IS NO LAUGHING MATTER!

1 New visor design has built-in antenna

2 Face print could show white hair or a layer of ice!

3 Standard breathing apparatus has not changed since 1978!

4 Small Ice Planet logo on torso

ICE DRIVE

The Ice Planet Chief features in two circuits of the LEGO *Racers* video game.

STONE COLD CLASSIC!

HEAD
TENTACLES

ALIEN TROOPER

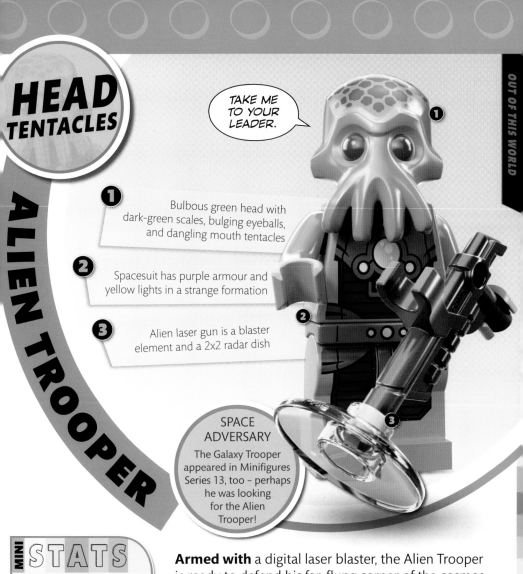

TAKE ME TO YOUR LEADER.

❶ Bulbous green head with dark-green scales, bulging eyeballs, and dangling mouth tentacles

❷ Spacesuit has purple armour and yellow lights in a strange formation

❸ Alien laser gun is a blaster element and a 2x2 radar dish

SPACE ADVERSARY
The Galaxy Trooper appeared in Minifigures Series 13, too – perhaps he was looking for the Alien Trooper!

MINI STATS

Theme
LEGO® Minifigures

Year
2015

First appearance
LEGO Minifigures Series 13

Rarity

Armed with a digital laser blaster, the Alien Trooper is ready to defend his far-flung corner of the cosmos. If his laser gun doesn't prove startling enough, his pulsating tentacled head will! Its lime-green design was created especially for this Minifigure.

TICKLED BY TENTACLES
The Alien Trooper's head mould is a joke prop in 2019's People Pack – Space Research and Development (set 60230). An astronaut shocks his co-workers with an alien mask on a stick, clearly amusing himself more than anyone else!

51

SCIENCE-FICTION FIGURE

SEEN ANY FLYING SAUCERS AROUND HERE?

1 New open-fronted helmet mould with bright-orange earpieces

2 Metallic space-armour collar with rivets

3 Sand-green spacesuit with orange lightning bolts

ATOMIC ARMS

The Retro Space Hero's upper arms are printed with atoms – an iconic symbol of science fiction from the early 1900s.

The Retro Space Hero landed in the LEGO Minifigures theme straight out of science fiction! An old-school space adventurer, he is dressed in an all-new sand-green spacesuit, vibrant cape, and yellow-crested helmet – fully prepared to face any intergalactic threats thrown at him.

FELLOW SPACE FANATIC

Rocket Boy is more rocket than boy – his costume is bigger than he is! Like the Retro Space Hero, he dreams of out-of-this world adventures.

MINI STATS

Theme
LEGO® Minifigures

Years
2017

First appearance
LEGO Minifigures Series 17

Rarity

Theme
LEGO® Minifigures

Years
2019

First appearance
LEGO Minifigures
Series 19

Rarity

According to the poster the Galactic Bounty Hunter is holding, he is looking for a LEGO bounty – built or broken! In his all-black armour and identity-concealing helmet, he cuts a menacing figure as he hunts down targets in the outer limits of space.

BLACKTRON
REBORN

WHO IS HE? Anonymity is key for the Galactic Bounty Hunter, but removing his helmet reveals a blue alien face.

GALACTIC BOUNTY HUNTER

I KNOW I LOOK GREAT, BUT LOOK AT THE POSTER!

1 Helmet mould shared by Ant-Man in LEGO® Marvel™ Super Heroes

2 Chest armour displays the Blacktron I logo – a faction of bounty hunters in LEGO® Space from 1987

3 Dark grey and copper leg armour

WANTED
BUILT or BROKEN
REWARD
#300,000,000

BACK IN FACTION
Another LEGO Space blast from the past appeared in Minifigures Series 13. The Galaxy Trooper's heavy armour bears the insignia of Galaxy Squad, a faction of space soldiers from 2013.

53

This minifigure is one of two astronauts, two crew members, and a robot stationed aboard the Lunar Space Station. He is wearing a white-and-orange Extra-Vehicular Mobility suit – a new style for LEGO City minifigures in 2019 – with a ready supply of oxygen on his back. It allows him to move freely outside the modular station that he calls home, monitoring it and making repairs.

MINI STATS

Theme
LEGO® City

Year
2019

First appearance
Lunar Space Station (60227)

Rarity

ASTRONAUT

COOL SPACE SUIT

WANT TO SEE MY MOONWALK?

1 Reflective sun visor in metallic gold

2 Combined helmet and oxygen tank element also used by divers in LEGO® Atlantis sets

3 Torso features a pressure valve and LEGO City Explorers logo under helmet

4 Orange-trimmed pressure suit printing continues on legs

SPOT THE SATELLITE

This astronaut is one of two crew members at work inside the Lunar Space Station. Her practical and comfortable blue jumpsuit has a white satellite logo on one side.

BIG PINK BRAIN!

THIS GALAXY IS MINE!

1 Transparent pink brain is integral to green head piece

2 Chunky alien ray gun

3 Eyebrows, eyes, and lips printed on head piece

4 Printed slope piece instead of who knows how many legs!

ALIEN VILLAINESS

Don't mess with the Alien Villainess! She's dressed to intimidate in her black and pink dress, and purple cape with a pointed collar, which is made from two separate parts. Her alien head also has a pink brain, rather than the more "usual" green kind.

ALIEN INVASION
The Alien Villainess looks remarkably similar to the Alien Commander from the LEGO® Alien Conquest theme. Perhaps they intend to conquer the galaxy together?

MINI STATS

Theme
LEGO® Minifigures

Year
2012

First appearance
LEGO Minifigures Series 8

Rarity

55

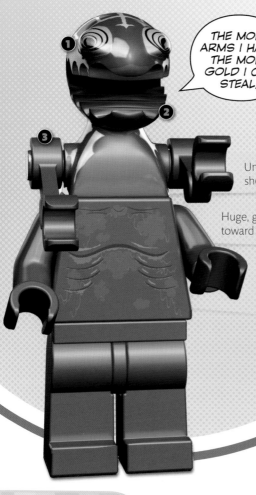

THE MORE ARMS I HAVE, THE MORE GOLD I CAN STEAL!

FOUR-ARMED ALIEN!

① Unique head and shoulder piece

② Huge, gaping mouth extends toward back of head

③ Extra limbs use same mould as LEGO® *Star Wars*™ battle droid arms

FRENZY

MINI STATS

Theme
LEGO® Space

Year
2009

First appearance
Gold Heist (5971)

Rarity

Very definitely armed and dangerous, this space bandit is on the most-wanted list – both in the Space Police III subtheme and among minifigure collectors. If you spot his unique head peeking out of a box of LEGO bricks, expect a Frenzy!

MULTIPLE PLAYERS

Frenzy isn't the only LEGO minifigure with four arms. There are several others, including General Kozu (pictured) and Pong Krell.

MINI STATS

Theme
LEGO® Space

Years
1991–1992

First appearance
Grid Trekkor (6812)

Rarity

Rocketing into the LEGO Space theme in 1991, the Blacktron II Commander is an update of the original Blacktron minifigures from 1987. More colourful than his all-black predecessor, his enhancements include a jetpack accessory with twin handles.

THE NEW BLACK!

BLACKTRON II COMMANDER

I'M JOINING THE JET SET!

1 Same helmet and visor combo as M:Tron minifigures

2 New jetpack piece fits over torso

3 New Blacktron logo also worn by Blacktron Fan in THE LEGO® MOVIE™

COME FLY WITH ME

1991 also saw the release of an M:Tron minifigure with a classic jetpack element, first seen in 1985.

Theme
LEGOLAND® Space

Years
1978–1986

First appearance
Rocket Launcher (462)

Rarity

The red spaceman and his fellow explorers signified a giant leap for LEGO minifigures. Boldly going where no bricks had gone before, they were the first to blast off into LEGOLAND Space. They also pioneered the visor-free helmet, which appeared in sets until 1988.

SUITS YOU!

The classic red spaceman was the first to shoot for the stars, but the white, yellow, blue, and black spacemen soon followed him. The black spaceman is the rarest of them, appearing in eight sets.

SPACEMAN

UNDER
THE HELMET

In THE LEGO® MOVIE™ featurette *Behind The Bricks*, the red spaceman can be glimpsed without his helmet, revealing a slick black hairdo!

RED ALERT!
RED ALERT!

1 Original helmet mould was also used in LEGOLAND Town and Castle sets

2 Oxygen tanks attached with a neck bracket

3 Iconic LEGO® Space logo is still used today

FIRST
MINIFIGURE
IN SPACE

EIGHT-EYED ALIEN TOUGH GUY!

SNAKE

> I'M KEEPING MY EYES ON YOU!

1 Spiked helmet is open at the back

2 Standard black visor can hide face – thankfully!

3 Head piece has seven eyes and four big fangs

4 Eighth eye in middle of torso

MINI STATS

Theme
LEGO® Space

Year
2009

First appearance
Space Truck Getaway
(5972)

Rarity

Snake is the most prolific bad guy in the Space Police III subtheme, featuring in four sets. His helmet has a movable visor that's perfect for hiding his identity. Which is just as well, as there's no mistaking his beady eyes – all eight of them!

GANGING UP

Snake is a member of the Black Hole Gang. This band of bad eggs is made up of criminals from the Space Police III subtheme, including Squidman, Frenzy, Kranxx, and the terrible Skull twins!

BARE FACED

Snake is identical in all his sets, except Space Speeder (set 8400), where he has no visor.

WE COME IN PIECES!

1 Helmet features an intricate wire pattern

2 Transparent red head piece beneath helmet is printed with menacing eyes

3 Bulky armour fits under head and adds height

ALPHA DRACONIS

WRITTEN IN THE STARS
The name of a real star, Alpha Draconis means "head of the dragon".

A major antagonist in the UFO subtheme, Alpha Draconis was one of the very first alien minifigures. This intergalactic baddie has an elaborate helmet, which you can remove – if you dare! Underneath is a head only a mother(ship) could love.

ALIEN INVASION
Alpha Draconis is the head of a race of aliens from the planet Humorless. These strange-looking, mind-reading beings come in red and blue variations.

MINI STATS

Theme
LEGO® Space

Year
1997

First appearance
Alien Avenger
(6975)

Rarity

MINI STATS

Theme
LEGO® Minifigures

Year
2014

First appearance
LEGO Minifigures
Series 12

Rarity

Not only does the Space Miner have a cool orange drill, he also has a new helmet and face print. His armour has been seen before, though: on Galaxy Patrol from Minifigures Series 7, Alien Avenger from Series 9, Lex Luthor from LEGO® DC Comics Super Heroes, and Infearno from LEGO® Ultra Agents.

LICENSE TO DRILL!

MINER DETAIL
The LEGO® Space logo is hidden on the miner's armour, but with a drill in place of the rocket.

SPACE MINER

ALL BACK TO MINE!

"The colour scheme went back and forth, but in the end we used almost the same colours as Luis Castañeda's concept sketch."
CHRIS B. JOHANSEN,
LEGO DESIGN MASTER

❶ New helmet with clear visor

❷ Orange ray gun element with detachable drill bit

❸ Detailed printing includes steel toe caps

61

This robot has legs! Not just any old legs, either, but the very first printed legs and hips seen on a LEGO minifigure. An automated underling of the villainous forces of Spyrius (which gave the LEGO Space Spyrius subtheme its name), he also boasts a new robot face print, and a totally clear helmet with no visor – another minifigure first!

MINI **STATS**

Theme
LEGO® Space

Year
1994

First appearance
Lunar Launch Site
(6959)

Rarity

SPYRIUS DROID

I'M ALL WIRED OUT!

①

① New transparent version of standard LEGO helmet

② Head piece also used for a droid in the 1996 theme LEGO® Time Cruisers

③ First-ever printed legs and hips

②

③

I SPY

The Spyrius Droid can be seen in the LEGO.com video *H.Q. Briefing*, where he wears a Space Police I uniform.

RISE OF THE ROBOTS

Spyrius was the first subtheme to cast minifigures as robots.

FIRST PRINTED LEGS

CHAPTER FOUR
OU'RE MY
HERO!

MEET THE
MINIFIGURES
WHO WILL FIGHT
FOR WHAT'S RIGHT
AND PROTECT US
ALL FROM THE
BAD GUYS!

KAI (ELEMENTAL ROBES)

"NINJA, GO!"

1 Same head wrap as Kai's ZX variant

2 Standard Kai face print with distinctive left-eye scar

3 Ornate black kimono with red and gold pattern

NINJA WEAPON
This Kai variant wields a double-edged Fire blade.

The LEGO NINJAGO Ninja of Fire dons his elemental robes for the first and only time in Kai's Fire Mech. Kai's formal kimono is mostly black – unlike his original red robes – with the Ninja's colour used only for detail. The printing continues on the back of his kimono, with a large Fire symbol.

MINI STATS

Theme
LEGO® NINJAGO®

Year
2013

First appearance
Kai's Fire Mech (70500)

Rarity

TAKING TO THE AIR
The Rattlecopter (set 9443) sees the ZX variant of Kai blasting into battle courtesy of an awesome rocket pack!

All hail Queen Halbert of Knighton! This royal ruler of the LEGO NEXO KNIGHTS theme is more than a figurehead – she is also a brave warrior. The gold-and-silver armour printing on this battle-ready variant of her is fit only for a queen, as it is unique to her. She appears in just one other set, in her royal robes instead of armour.

ROYAL KNIGHT

QUEEN HALBERT

> MINIFIGURES CAN BOW, AND MY ENEMIES WILL DO SO!

1 Crown-braid hairstyle, first worn by Princess Leia in LEGO *Star Wars*™ sets

2 Distinctive beauty mark and calm smile

3 Printed dragon symbol can be seen through hole in armour

4 Blue clothing beneath metallic armour

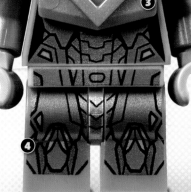

STONE STATUE

There is a white stone statue of Queen Halbert in Knighton Castle (set 70357). Her head, torso, and slope skirt show cracks in the stone.

The Rock Raiders would be lost without their skilled and strong-willed pilot, Flight Lieutenant Jet. The sole female member of this team of space travellers has appeared in six Rock Raiders sets, and as a minifigure keychain.

SHE ROCKS!

MINI  **STATS**

Theme
LEGO® Rock Raiders
Year
1999–2000
First appearance
Hover Scout (4910)

Rarity

JET

ROCK READERS

The first Rock Raiders sets all included special *Rock Raiders* comics.

I'M ALWAYS IN CONTROL!

1 Helmet is identical to those worn by LEGO Exploriens

2 Blonde hair and headset printed on head piece

3 Grey harness and pouch on torso

MINE CAST

Jet's fellow Raiders are driver Axle, helmsman Bandit, commander Chief, geologist and explosives expert Docs, and mechanic Sparks.

NINJA BIKER

HOW WILL I FIT A BIKER HELMET OVER THIS HAIR?

1

ZANE (SNAKE JAGUAR)

1 White flat-top hair piece, specially designed for Zane

2 Black bandana – Zane usually wears one in white

3 Colourful zipped-up biker vest

BIKER FASHION
Several of the Sons of Garmadon gang wear the same biker vest as Zane, which has "S.O.G." insignia on the back.

As the Ninja of Ice,
Zane is usually in white Ninja robes. He adopts this tough-guy disguise to pose as Snake Jaguar – made-up member of the criminal biker gang Sons of Garmadon. "Snake" appears in just one LEGO set, when he infiltrates their headquarters!

MINI STATS

Theme
LEGO® NINJAGO®

Year
2018

First appearance
S.O.G. Headquarters (70640)

Rarity

BAD BOYS

Zane is just posing as a villainous biker, but these guys are the real deal. Nails (left) and Skip Vicious hang out with "Snake" in the S.O.G. Headquarters set.

BOW DOWN TO ME OR SUFFER MY WRATH!

1 Moulded hood over short fabric cape

2 Head can be turned around to reveal a worried look

3 Green "5" on torso is a clue to Lloyd's future as the fifth Ninja

4 Shorter legs seen only on this version of the character

SON OF THE DARK LORD

Lloyd is the son of LEGO NINJAGO villain Lord Garmadon, but he grows up to be a good guy and fights against evil as the brave Green Ninja!

GOLDEN BOY

When Lloyd becomes the Ultimate Spinjitzu master, his robes turn from green to gold! This is one of two Golden Ninja variants of Lloyd.

MINI STATS

Theme
LEGO® NINJAGO®

Year
2012

First appearance
Rattlecopter
(9443)

Rarity

MINI STATS

Theme
LEGO® Hidden Side

Year
2019

First appearance
Wrecked Shrimp Boat
(70419)

Rarity

By day, Parker L. Jackson is a high-school student. By night, she hunts down the ghosts of her haunted hometown! This LEGO Hidden Side hero has a new combined beanie-and-plaits hair piece and a double-sided head with a suspicious, spook-spotting alternative expression.

A REAL GHOUL-GETTER

PARKER L. JACKSON

I'M ON A WILD GHOST CHASE!

1 Yellow beanie and lavender plaits

2 Cross-body bag and "P" pendant necklace printed on torso

3 Ripped jeans

LET LOOSE
This variant of Parker first appeared in Jack's Beach Buggy (set 70428) in 2020. She has a brand-new hair mould, with loose lavender curls and a pair of built-in headphones.

If there's a newsworthy event in LEGO City, this minifigure is there to report it. Dressed in a trim red jacket and proudly displaying her press badge, she is ready to do a piece to camera. Her first assignment was pretty sweet – she covered the opening of a new donut shop in LEGO City!

LEGO PRESS PASS

Theme
LEGO® City
Year
2019
First appearance
Donut Shop Opening (60233)

Rarity

REPORTER

DONUT MESS THIS UP!

NEWS IDOL

LEGO® Space news reporter Lotta Brix from the Alien Conquest subtheme sported a similar red blazer in 2011.

1
Reddish-brown hair swept back into a tidy twist at the back

2
Pinstriped blazer with a minifigure press badge clipped onto the collar

3
Neat black trousers

PRESS

READY TO ROLL

Behind the camera, the more casually dressed Cameraman captures the "hole" story in the Donut Shop Opening set.

EL FUEGO

1 Flame design on back of helmet (El Fuego means "The Fire" in Spanish)

2 Plaster – being a stuntman is a dangerous business!

3 Target-motif T-shirt under his overalls

4 Star-patterned cleaning cloth in pocket

GHOST-HUNTING STUNTMAN

El Fuego is every bit as fearless and heroic as his red cape and target helmet suggest! The ID badge on his oil-stained green overalls is for his day job as a school groundskeeper. By night, he is the most daring member of the LEGO Hidden Side ghost-hunting team.

DOUGLAS ELTON

In his day job, El Fuego goes by his real name, Douglas Elton. He first appeared as Douglas, with slicked-back hair, in J.B.'s Ghost Lab (set 70418).

MINI STATS

Theme
LEGO® Hidden Side

Year
2019–2020

First appearance
El Fuego's Stunt Truck (70421)

Rarity

71

A KEY CHARACTER!

IS THIS A WIND-UP?

1 Hat mould created for LEGO® Pirates theme in 1989

2 Chin strap is printed on head

3 Key fits onto red neck bracket

4 Gold printing on arms and legs

TOY SOLDIER

Theme
LEGO® Minifigures

Year
2013

First appearance
DK's LEGO Minifigures Character Encyclopedia book

Rarity

This rosy-cheeked fellow is found exclusively as part of the DK LEGO *Minifigures Character Encyclopedia* from 2013. He features a rare (and sadly non-functional!) wind-up key piece, and was the first character in the LEGO Minifigures theme not to be part of a wider series.

MILITARY MATES

The Toy Soldier shares his "shako" cap style with the Imperial Soldiers from LEGO Pirates. These brave souls battle against Captain Redbeard and his scurvy crew!

MINI STATS

Theme
LEGO® Minifigures

Year
2017

First appearance
LEGO Minifigures
Series 17

Rarity

The ethereal Elf Maiden has moved beyond her Elvish homeland to seek adventures in the LEGO Minifigures theme. Should she meet any enemies, her gold shield and sword are poised for action. Her silver-sleeved, leaf-patterned dress is printed with reflective paint so it shimmers in the light.

EARLIER ELF
A male Elf dressed in a green cape appeared in Minifigures Series 3 in 2011. Perhaps the Elf Maiden is his friend or kin.

ELF MAIDEN

ELVISH ELEGANCE

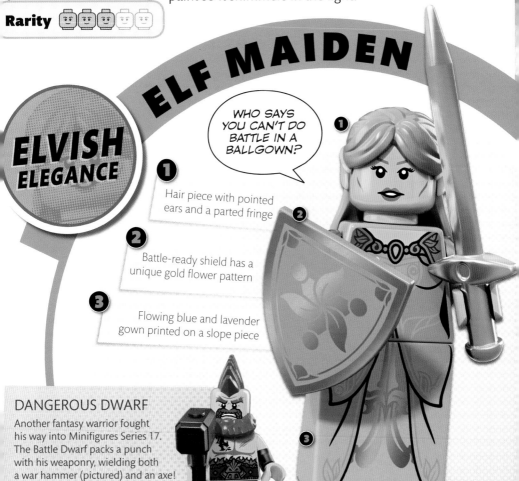

WHO SAYS YOU CAN'T DO BATTLE IN A BALLGOWN?

1 Hair piece with pointed ears and a parted fringe

2 Battle-ready shield has a unique gold flower pattern

3 Flowing blue and lavender gown printed on a slope piece

DANGEROUS DWARF
Another fantasy warrior fought his way into Minifigures Series 17. The Battle Dwarf packs a punch with his weaponry, wielding both a war hammer (pictured) and an axe!

73

This **hoodie-wearing** ghost hunter is a regular in the LEGO Hidden Side theme. There are several variations of him in different clothing, but this 2019 version was the first. He is never seen in sets without his ghost dog, Spencer, by his side!

GHOST DOG OWNER

MINI **STATS**

Theme
LEGO® Hidden Side

Year
2019

First appearance
J.B.'s Ghost Lab (70418)

Rarity

JACK DAVIDS

CHECK OUT MY BOO-TUBE CHANNEL!

BARK FROM THE DEAD

Jack's beloved pet dog, Spencer, died in a car crash but returned as a ghost dog, with a marbled white and transparent body.

1 Combined hood and cap piece was specially created for Jack

2 Red letterman jacket with white hoodie underneath

3 Dog lead wrapped around waist

MUMIFIGURE

Jack's mother, Rose Davids, is also a teacher at his high school. Dressed in a sharp suit and pearls, she appears in the Newbury Haunted High School (set 70425) and transforms into a Gloombie. Better call Jack!

FLYIN' LION!

LAVAL

Speech bubble: 3... 2... 1... WE HAVE LIFTOFF!

1 Lion headgear fits over standard minifigure head mould

2 Eye holes show printing on head piece

3 Multipart jetpack fits onto armour

4 Unique printed armour on legs has lion detail

MINI STATS

Theme
LEGO® Legends of Chima™

Year
2013

First appearance
Scorm's Scorpion Stinger (70132)

Rarity

Laval, Prince of the Lion Tribe of Chima, comes in many variants, but only one has an awesome rocket-powered jetpack! The jetpack has eight parts, including two LEGO® *Star Wars®* lightsaber hilts, and appears in Scorm's Scorpion Stinger.

> "This minifigure was a challenge! Especially the element design – we had so many technical constraints."
> ALEXANDRE BOUDON, LEGO DESIGN MASTER

OUTLANDS DEFENDER

Laval appears in the same armour, minus the jetpack, while he's defending Lavertus's Outland Base (set 70134) from rival Tribes.

HAS ANYONE SEEN MY RAZOR?

TEAM PLAYERS

Dash Justice and his team also appeared in their own video game.

DASH JUSTICE

1 New flat-top hair piece, later worn by Zane in LEGO® NINJAGO® sets

2 Headset detail on unique face print

3 Detailed torso print with Alpha Team logo

MINI STATS

Theme
LEGO® Alpha Team

Year
2001

First appearance
Alpha Team Helicopter (6773)

Rarity

Minifigures don't come much cooler than Dash Justice – secret agent and leader of the fearless Alpha Team. Dash has appeared in multiple variants, with different looks for many exciting missions. His detailed black suit is perfect for stealth operations.

DASH AND ALL

The Deep Sea variant of Dash has a diving helmet, while the stubble-free Deep Freeze variant dons fetching shades. However, they both sport Dash's trademark lopsided grin!

MINI STATS

Theme
LEGO® NINJAGO®

Year
2011

First appearance
Nya (2172)

Rarity

Formidable warrior Nya is the sister of Kai, the Ninja of Fire. She was the first female minifigure in the LEGO NINJAGO theme, and one of the first NINJAGO minifigures to have a reversible head. Like Kai, Nya trains as a Ninja and becomes the powerful Ninja of Water.

FIRST LADY OF NINJAGO ISLAND!

NYA

I'M NOT LETTING THE BOYS HAVE ALL THE FUN!

 1 Hair piece first seen on Irina Spalko in LEGO® Indiana Jones™

 **2** Reverse of head piece shows Ninja mask

 **3** Red robes with phoenix print echo those of her brother, Kai

WHOLE NEW WORLD

Nya has a cameo in THE LEGO® MOVIE™, as Wyldstyle is telling Emmet about all the realms in the LEGO universe.

THE X FACTOR

The Samurai X version of Nya has a helmet similar to Lord Garmadon's, but with a chin guard that covers her mouth. Underneath is another double-sided head – determined on one side, angry on the other.

Theme
LEGO® Minifigures

Year
2019

First appearance
LEGO Minifigures
Series 19

Rarity

This intrepid adventurer is equipped with everything he could ever need in the jungle. His wide-brimmed hat will keep away the sun's rays and jungle critters. His brand-new backpack holds his bedroll. And his working magnifying glass lets him peer at a new species of chameleon – first found by his side in the Minifigures line!

BACK TO NATURE

JUNGLE EXPLORER

1 Backpack and bedroll piece attaches at his neck

2 Torso features a handy rope slung over his field shirt

3 Bright green and orange-striped chameleon first spotted in 2019!

RETRO RESEMBLENCE

The Jungle Explorer is an homage to Johnny Thunder, who starred in LEGO® Adventurers sets from 1998.

URBAN JUNGLE

This explorer is one half of a daring duo who first appeared in jungle-themed LEGO® City sets in 2019. She takes on a crocodile in the Jungle Starter Set (set 60157).

CHAPTER FIVE
SPOOKY AND SCARY

BEWARE!
YOU ARE ABOUT
TO OPEN A VAULT
OF MINIFIGURE
FIENDS! DARE
YOU TURN
THE PAGE?

This smiling spirit proved very popular upon its 1990 release – and appeared in seven sets over five years, haunting the Black Knights, the Royal Knights, and even an old tree! Beneath the Ghost's new shroud piece is simply a white torso and a plain black head.

GLOWS IN THE DARK!

"I designed and sculpted the Ghost. In those days, that meant making a wooden prototype at five times actual size!"
NIELS MILAN PEDERSEN, LEGO DESIGNER

Theme
LEGO® Castle

Years
1990, 1992–1993, 1995

First appearance
Black Monarch's Ghost (6034)

Rarity

GHOST

WOOOO!

1 Ghoulish face still has a classic LEGO smile!

2 Plain black head visible through mouth and eye holes

3 Glow-in-the-dark ghost piece slots over head and torso

4 1x2 brick and 1x2 plate instead of legs

1
2
3
4

CHILLING CHAP

This sunken-eyed Spectre haunted Series 14 of the Minifigures theme. His ghostly form hovers on a whirling transparent base.

HEAD BOLTS TOGETHER

AARGH! A GHOST!

1 Unique head extension has LEGO studs on both sides

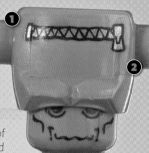

2 Zips printed on front and back of head extension, and on back of head

3 Exclusive patchwork jacket print on torso

THE MONSTER

TALL STORY
The Monster was originally going to have 1x1 plates on his feet to make him taller.

Fittingly, this classic monster comes with an extra body part – a unique head extension. This monstrous forehead is printed with zips at the front and back. Talk about getting down to the nuts and bolts!

MONSTERS, INC.!

The Monster isn't the only minifigure with a screw or two loose. The Monster from Minifigures Series 4 (pictured) and the Monster Butler from LEGO® Monster Fighters are variations of this lumbering villain.

MINI STATS

Theme
LEGO® Studios

Year
2002

First appearance
Scary Laboratory (1382)

Rarity

It is rare for a minifigure to have no hat or hair ❶

Dark stone-grey torso with medium stone-grey hands ❷

Drumstick is longer than a regular LEGO turkey-leg piece ❸

Shovel design is unchanged since 1979! ❹

FIRST ZOMBIE!

MINI STATS

Theme
LEGO® Minifigures

Year
2010

First appearance
LEGO Minifigures
Series 1

Rarity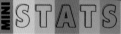

The first-ever zombie minifigure doesn't have much in the way of brains... He prefers a turkey leg! His inclusion in the first series of collectible Minifigures confirmed the eclectic nature of the theme, and pushed the boundaries of what a minifigure could be.

HALLOWEEN HORROR

The Zombie also appears in Halloween Accessory Set (set 850487), but with a brown suit and a grey tie.

IT'S CATCHING!

Minifigures Series 1 also introduced the Skateboarder, who was later reimagined as a zombie.

MINI STATS

Theme
LEGO® NINJAGO®
Year
2016
First appearance
Tiger Widow Island
(70604)

Rarity

When Earth Ninja Cole visits the haunted Temple of Airjitzu, he is turned into an eerie green ghost version of himself! Ghost Cole made a fleeting appearance in just two 2016 LEGO sets, before returning to his more human-like minifigure form.

WHO GLOWS THERE?

GHOST COLE

SORRY, DID I SPOOK YOU?

1 Tousled black hair

2 Ghostly green face has two expressions

3 Gold ghost emblem on "Skybound" Ninja robes

4 Grey knee straps

GHOST NINJA

With glowing transparent limbs, Morro is a ghost Ninja who leads a band of ghosts against his old teacher, Master Wu.

83

One of the henchmen of evil genius Dr Inferno, Slime Face just wants to recover sunken treasure – if only the LEGO Agents would leave him alone! The minifigure's head is coloured trans-neon green to make it look like slimy green jelly.

YUCKY-LOOKING VILLAIN!

MINI STATS

Theme
LEGO® Agents

Year
2008

First appearance
Deep Sea Quest
(8636)

Rarity

SLIME FACE

I'M SORRY, I HAVE A BAD COLD.

1 Rare helmet also worn by Mr. Freeze in LEGO® DC Comics Super Heroes

2 Transparent green head with slime printing

3 Dr Inferno's logo – a scary burning skull!

"The eyes and mouth really stand out. The one red eye is a vibrant contrast to the slime."
LAUGE DREWES, LEGO DESIGN MANAGER

CRUDE CREW

Slime Face isn't Dr Inferno's only mean-looking crony: Break Jaw and Gold Tooth also have uniquely evil appearances!

SPOOKY GIRL

1. Long, straight hair extends down front and back of torso

2. Spider emerging from pocket

3. Teddy bear element is also carried by the Panda Guy Minifigure – naturally his is black and white, too!

4. Grey fabric skirt

ALL **BLACK** AND **WHITE**

Spooky Girl is made of only black, white, and grey elements – the most noticeable being her new hair piece. She wears a black jacket and her legs are printed with stripy socks. She is the fifth collectible Minifigure to have a teddy bear, though only hers has one eye missing.

THE MONOCHROME SETS

Spooky Girl is the third monochrome collectible Minifigure. The first was the Mime from Series 2, and the second was the Sad Clown from Series 10.

MINI STATS

Theme
LEGO® Minifigures

Year
2014

First appearance
LEGO Minifigures Series 12

Rarity

FANCY A BITE?

1 New slicked-back hair piece

2 Head can be turned around to reveal an open mouth

3 Flowing cape is red on one side, brown on the other

VAMPIRE

HAIR HEIR
The Vampire's hair has since been worn by Wolverine from LEGO® Marvel Super Heroes.

This caped count has developed quite a "fang" club over the years. Not only is he the original vampire minifigure, he's also the first to have a combed widow's peak hair piece. His double-sided head shows either his toothy smile, or him opening his mouth in anticipation of a juicy neck.

BITE-SIZED FACT
Slay one vampire and another takes its place. Thankfully, the Vampire from the second series of collectible Minifigures is far from your average bloodsucker, favouring a fruit smoothie to a pint of the red stuff.

MINI STATS

Theme
LEGO® Minifigures

Year
2015

First appearance
LEGO Minifigures
Series 14

Rarity

The **Minifigures theme** has regularly featured characters dressed up in strange costumes. In the past we've seen minifigures dressed as bunnies, chickens, gorillas – and even hot dogs. But a guy dressed as a man-eating plant? Now that's a first!

BEYOND BE-LEAF!

PLANT MONSTER

EAT YOUR GREENS – BEFORE THEY EAT YOU!

1 Terrified expression – is this monster scared of itself?

2 Unique carnivorous plant piece surrounds head

3 Dark-green legs and torso are the same colour as most LEGO trees and plants

WHY WORRY?

The Plant Monster is not the first character in the Minifigures line to look worried. Hazmat Guy has a similar expression, despite his protective suit!

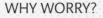

87

TACKLE THE JACKALS

Helping the Flying Mummy to guard Scorpion Pyramid (set 7327) are two Anubis Guards. These jackal-headed warriors each carry a sword and a scarab shield.

Unearthed in 2011 as part of the Pharaoh's Quest theme, the Flying Mummy has a truly extravagant wing piece. It attaches to the minifigure's neck and features 22 tan feathers and 18 blue ones.

MINI STATS

Theme
LEGO® Pharaoh's Quest

Year
2011

First appearance
Flying Mummy Attack (7307)

Rarity

FLYING MUMMY

1
Falcon headdress resembles Ancient Egyptian god Horus

2
Reverse of head shows just one unbandaged eye

1

2

COME FLY WITH ME! MUMMY BACK GUARANTEED!

3
Elaborate printed wings attach with a neck bracket over the minifigure's neck

3

SPECTACULAR WING PIECE

DAWN OF THE WED!

I DON'T MEAN TO MOAN, BUT...

1 No other minifigure wears a ponytail in this tan colour

CORDON BLEURGH!
The Monster Fighters theme also includes a Zombie Chef.

2 Reversible head shows open mouth on other side

3 Unique torn bridal gown print on slope piece and torso

ZOMBIE BRIDE

On her big day, the Zombie Bride is dressed to digest in her unusual sloped piece, printed like a decayed wedding gown. Minifigures should be careful if they receive an invitation to her wedding – it might just be them on the menu...

LOVE AT FIRST FRIGHT

The Zombies set is well named, featuring not only a Zombie Bride, but also a top-hatted Zombie Groom, and a Zombie Driver in a tattered chauffeur's uniform.

MINI STATS

Theme
LEGO® Monster Fighters

Year
2012

First appearance
The Zombies (9465)

Rarity

DON'T TELL ANYONE, BUT I CAN'T SWIM!

BLING KING
The Emperor's breastplate is also worn by 2011's City of Atlantis Golden King.

① Helmet hides fishy facial features!

② Armour and helmet are black with gold printing

③ Portal key is one element

PORTAL EMPEROR

RUST IN PEACE!

One of the six guardian races of the underwater world of Atlantis, the Portal Emperor includes a new helmet with a never-before-seen gold and black speckled paint pattern. This minifigure is made of five parts, four of which are unique.

PORTAL KEYS
All but two of the LEGO Atlantis sets come with specially moulded portal keys, which were the object of the divers' quest in the 2010 storyline.

MINI STATS

Theme
LEGO® Atlantis

Year
2010

First appearance
Portal of Atlantis (8078)

Rarity

MINI STATS

Theme
LEGO® Hidden Side

Year
2019

First appearance
Shrimp Shack Attack
(70422)

Rarity

When this shellfish chef's Shrimp Shack is attacked by ghosts, he becomes possessed! He has not only turned a gruesome shade of green – he has also grown in size with a torso extender piece. Chef Enzo also has a friendlier head for before and (hopefully) after his possession.

SHRIMPLY SCARY

CHEF ENZO (POSSESSED)

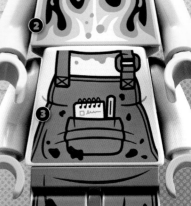

① Food service cap – even as a ghost, he remembers hygiene!

② Swirling "Gloombie" energy on torso and face

③ Grease-stained apron with notepad and pen in pocket

LIVE GHOSTS!
Hidden Side was the first LEGO theme to combine building with a digital app that brings the ghosts in the haunted sets to life.

GHOST SHIFT
Chef Enzo isn't the only Shrimp Shack staff member to become possessed. Sally, the server, also turns into a Gloombie (without dropping her tray!).

NO BODY COMES CLOSE!

Trick or treat! This spooky fellow is all ready for the scariest night of the year. Skeleton Guy pays tribute to the enduring LEGO Skeleton figure, but this time in actual minifigure form. He is especially notable as one of the earliest minifigures to have printing on his arms and the sides of his legs.

MINI STATS

Theme
LEGO® Minifigures

Year
2015

First appearence
LEGO Minifigures
Series 14

Rarity 😀 😀 😀 😐 😐

SKELETON GUY

I'D LIKE TO SCARE PEOPLE BUT I DON'T HAVE THE GUTS!

1 Skeleton mask's side strings go around the head and tie at the back

2 Printing designed to look like LEGO Skeleton figures

3 Creepy pumpkin basket is the same mould as the Leprechaun's clay pot

"Some years ago, I made a drawing of how I thought Skeleton Guy should look, and apart from the eyes, the design didn't change much."
CHRIS B. JOHANSEN, LEGO DESIGN MASTER

SPECIAL UNDEAD VARIANT

UURR... MUST... DO... OLLIE!

ZOMBIE SKATEBOARDER

① Same expression as Series 1 Skateboarder

② Skull print now has "dead" X-shaped eyes

③ Studs hold minifigure on skateboard

The first Skateboarder

Minifigure was part of Minifigures Series 1, but this zombie version of him came with the *I Love That Minifigure* book from DK. His sunken red eyes and grey skin colour mark him out as part of a long line of LEGO zombies that started in Series 1.

"I always like it when we reimagine existing characters, so we can see what happened to them. I guess this guy wasn't such a good skater in the end..."
CHRIS B. JOHANSEN, LEGO DESIGN MASTER

MINI STATS

Theme
LEGO® Minifigures

Year
2015

First appearance
DK's *I Love That Minifigure* book

Rarity

93

I KNOW YOUR EVERY MOVE...

1 Long, blood-red hairpiece

2 Same face as the chess set's Evil Bishop

3 Skeleton armour resembles ribs

4 Plain black slope instead of legs

TERRIFYING **CHESS PIECE!**

EVIL QUEEN

Released as part of the Fantasy Era Castle Chess Set, the Evil Queen, alongside the Evil Wizard, leads a skeleton army against the Crown King. She has long hair that covers a red-eyed skeletal head, and a torso piece featuring armour with a skull motif.

WITCH SWITCH

The Castle Giant Chess Set (set 852293) from 2008 replaced the Evil Queen with an Evil Witch, who was also in the 2008 LEGO Castle Advent Calendar (set 7979), so the Evil Queen is rarer!

MINI

Theme
LEGO® Minifigures

Year
2018

First appearance
LEGO Minifigures
Series 18

Rarity

This monochrome Minifigure loves to hide in a corner at parties, then jump out at people! His suit has an eight-limbed arachnid body that attaches at his neck, and a fanged helmet that adds four eyes to his original two.

SPOOKY SPIDER

SPIDER SUIT BOY

I FOUND THIS COSTUME ON THE WEB!

1 Startlingly white head has a scared expression on other side

2 Zipped up spider-web hoodie

3 Spider's thorax, abdomen, and eight legs are made of rubber

FAMILIAR FANG

Spider Suit Boy looks spookily like Spooky Boy, who appeared in Series 16 of the LEGO Minifigures theme with a matching fang.

Spooky Tales

95

SCARY SKULL!

This frightening-looking minifigure sends shivers down your spine! A variant of the Skeleton Drone that first appeared in LEGO Alpha Team in 2002, the Super Ice Drone has a unique black skull head piece with white printing and red eyes, and its uniform features a silver-lined scarab with a cross.

MINI STATS

Theme
LEGO® Alpha Team

Year
2005

First appearance
Scorpion Orb Launcher (4774)

Rarity

SUPER ICE DRONE

WHY THE COLD SHOULDER?

1 Unique scary face print

2 Transparent blue helmet exclusive to Alpha Team

3 Rare mismatched arm and leg colours

WHAT'S IN A NAME?

The Super Ice Drone is controlled by the villainous Ogel, whose name spelled backwards is LEGO. This bad guy was given this name because he represents the opposite of LEGO fun and play.

FLY MONSTER

HORRID HYBRID!

1 Two curved antennae allow him to taste and smell

2 Huge, hatched bug eyes and a proboscis for feeding

3 Segmented abdomen and tiny green hairs on torso and legs

4 First transparent-black wings

FLY ON FILM

The Fly Monster is inspired by the 1958 sci-fi movie *The Fly*, in which a scientist accidentally turns himself into a human fly!

Half-fly, half-minifigure, the Fly Monster was once a regular housefly. He emerged from a scientific experiment with an oversized bug head, wings, and one bright-red pincer hand! His pincer is also a crab-claw hand on pirate Davy Jones in LEGO® *Pirates of the Caribbean*™ sets.

WEIRD SCIENCE

Meet the scientist responsible for the Fly Monster. The Monster Scientist also appeared in Minifigures Series 14. Be very careful around his fly-printed flask!

MINI STATS

Theme
LEGO® Minifigures

Year
2015

First appearance
LEGO Minifigures Series 14

Rarity

97

1. Transparent green Gloombie head
2. Stormproof raincoat
3. Cosy black gloves

TWO BEARDS

Claus has not one but two new beard pieces. He has a bushy white beard when he isn't possessed.

CLAUS STORMWARD (POSSESSED)

LEGO Hidden Side's sea-hardy lighthouse keeper, Claus Stormward, has a very stormy night when he is possessed by a ghost! He has a windswept "Gloombie" beard and upturned bucket hat.

MINI STATS

Theme
LEGO® Hidden Side

Year
2020

First appearance
The Lighthouse of Darkness (70431)

Rarity

FISHERMAN FRIEND

Captain Jonas wears the same hat as Claus, in a classic fisherman's colour. He appears in the Wrecked Shrimp Boat (set 70419), where his possessed minifigure has tentacles!

MINI STATS

Theme
LEGO® Minifigures

Year
2019

First appearance
LEGO Minifigures
Series 19

Rarity

This knight-mare strode into the LEGO Minifigures theme straight out of LEGO history! He is based on the Fright Knights subtheme, which brought a creepy, supernatural edge to LEGO Castle sets from the 1990s. He has a ghostly pale head and a gruesome grin under his spooky transparent-blue plumed helmet.

VINTAGE VILLAIN

FRIGHT KNIGHT

I WORK THE KNIGHT SHIFT!

1. Classic knight helmet design in a new pearl dark-grey shade

2. Armoured breastplate has a copper bat-wing pattern

3. Wide sword in eerie transparent blue

4. Shield printed with the Fright Knights emblem

FANTASY FOE

Another classic LEGO® Castle antagonist is the evil wizard character. A LEGO Minifigures version appeared in Series 13, in a blazing cloak and skull robes.

99

DANCING UNDEAD!

Her uniform might be ragged and rotting and her awkward grin may be missing a tooth, but the Zombie Cheerleader hasn't lost any of her school spirit! She waves her new green pom-poms with pride. She pleased the creepy crowds in the monsters-themed Series 14 of the Minifigures theme.

MINI STATS

Theme
LEGO® Minifigures

Year
2015

First appeerance
LEGO Minifigures
Series 14

Rarity

ZOMBIE CHEERLEADER

GIVE ME A Z...!

1
New bunches piece later seen on the Babysitter

2
Grey zombie face with red eyes and a dribbling grin!

3
Mouldy cheerleading vest with a "Z" for Zombie University

DAY OF THE UNDEAD

There was an outbreak of zombified characters in Minifigures Series 14. The Zombie Cheerleader was joined by a Zombie Pirate and a Zombie Businessman.

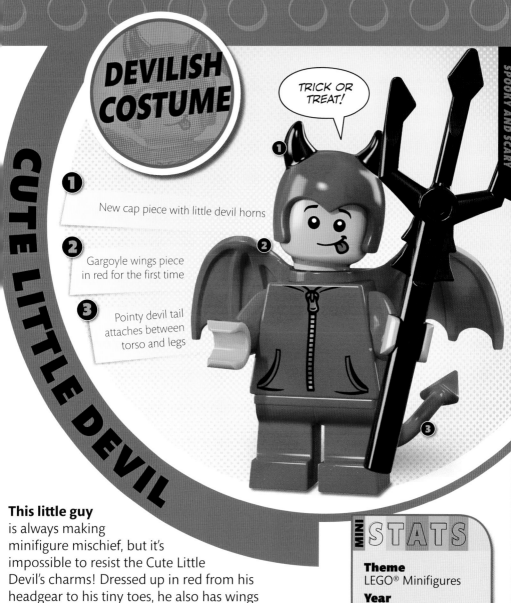

DEVILISH COSTUME

CUTE LITTLE DEVIL

TRICK OR TREAT!

1 New cap piece with little devil horns

2 Gargoyle wings piece in red for the first time

3 Pointy devil tail attaches between torso and legs

This little guy is always making minifigure mischief, but it's impossible to resist the Cute Little Devil's charms! Dressed up in red from his headgear to his tiny toes, he also has wings and a new pointy tail element. In his hand is a trident piece that doubles as a pitchfork.

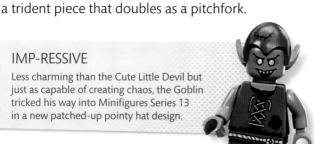

IMP-RESSIVE

Less charming than the Cute Little Devil but just as capable of creating chaos, the Goblin tricked his way into Minifigures Series 13 in a new patched-up pointy hat design.

MINI STATS

Theme
LEGO® Minifigures

Year
2016

First appearance
LEGO Minifigures Series 16

Rarity

BAD-NEWS BEARER

Beware of the Banshee!
This weeping and wailing Minifigure spirit is usually a harbinger of bad news. Many of her otherworldly features had not been seen in LEGO sets before she materialized, including her transparent hair – a minifigure first! – and her ghostly lower body, which was new in sand green.

BANSHEE

WAIT, IT MIGHT BE GOOD NEWS!

1 Wild, transparent-black wavy hair

2 Frayed bodice printing continues on her back

3 Wispy lower body first seen on ghosts in LEGO® NINJAGO® sets

MYTHICAL MYTH?

Another mythological Minifigure, Lady Cyclops from LEGO Minifigures Series 13, has her eye on anyone who believes that female Cyclopes don't really exist!

CHAPTER SIX
THE WORLD'S A STAGE

ROLL UP!
ROLL UP! PREPARE
TO BE AMAZED
BY THE SHOW-OFFS
AND ENTERTAINERS
OF THE LEGO
WORLD!

GIVE US A HISS!

1 New turban element

2 Rounded mouth, ready to play the pungi flute

3 Long shirttails printed on legs

4 New cobra accessory

SNAKE CHARMER

MINI STATS

Theme
LEGO® Minifigures

Year
2015

First appearance
LEGO Minifigures Series 13

Rarity

This collectible Minifigure is every bit the traditional Snake Charmer, with his white turban and waxed moustache. He comes with a pungi flute – and a rubber cobra, of course!

SSSLITHERING SSSSNAKESSS!

The Snake Charmer Mummy from LEGO® Pharaoh's Quest is another minifigure able to control snakes. Unfortunately, this bandaged baddie only uses his skills to cause trouble!

MINI STATS

Theme
LEGO® Minifigures

Year
2018

First appearance
LEGO Minifigures
Series 18

Rarity

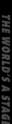

Need a laugh, a dance, or a balloon dog? The Party Clown is your guy! Bright and bubbly in both personality and sartorial style, this Minifigure lives to entertain. He was right at home in the party-themed Series 18 of the Minifigures line, celebrating 40 years of minifigures in 2018.

BALLOON ANIMALS

PARTY CLOWN

MAKE SURE YOUR PUP DOESN'T GO POP!

1 Two-tone top hat with purple flower print

2 Silly clown facepaint with a sticking-out tongue!

3 Bow tie piece in red for the first time

4 Fabric coat-tails attach above his hips

TALL ORDER
It looks like the Stilt Walker from the LEGO® City People Pack – Fun Fair (set 60234) has paid a visit to the Party Clown! He has a blue balloon dog. His stilts are standard 1x1x3 bricks.

105

The Pro-Surfer is about to take on some bodacious waves on his jawsome shark-printed surfboard – and perhaps some dolphins and fish in a race! He is the third surfer in the LEGO Minifigures line (the first appeared in Series 2), and just as laidback as his predecessors.

ICONIC HAIR

The Pro-Surfer shares a hair piece with Luke Skywalker from *LEGO® Star Wars™* sets.

MINI **STATS**

Theme
LEGO® Minifigures

Years
2017

First appearance
LEGO Minifigures
Series 17

Rarity

PRO-SURFER

> LIFE IS PRETTY SWELL.

1 Sun-bleached blond hair

2 Dark-blond eyebrows and stubble, with a relaxed smile

3 Red-and-black wetsuit with wave motif

SHARK SURFBOARD

SURF'S UP

Perhaps a good match for the Pro-Surfer in the waves, Surfer Girl from Series 4 has an identical surfboard piece with different printing.

NEW DRAGON DESIGN

> HEY, EVERYONE! CHECK OUT MY NINJA SKILLS!

1 Head wrap only reveals a determined frown

2 Unique torso and leg printing depicts Jay's lightning power

3 Back of torso has Jay's name in gold

2-DX
Jay DX appears in just two sets: Skeleton Bowling and Lightning Dragon Battle (set 2521).

JAY (DX)

The DX variant
of the LEGO NINJAGO
Ninja of Lightning has the same
head wrap and face as the original Jay
minifigure, but comes with sleek new torso and leg
printing. DX stands for Dragon eXtreme, so the design
features a dragon breathing lightning.

RAW ENERGY
The NRG Jay minifigure sees the NINJAGO
warrior transformed into pure lightning
energy. Talk about a bolt from the blue!

MINI STATS

Theme
LEGO® NINJAGO®

Year
2011

First appearance
Skeleton Bowling
(2519)

Rarity

107

CROSS MY PALM WITH SILVER STUDS.

1 Hair, purple bandana, and golden beads are all one element

2 Tower tarot card is printed on a 1x2 tile

3 Dark-red slope piece with detailed printing

FORTUNE-TELLER

LEGO TAROT CARDS

Every element that makes up the Fortune-Teller Minifigure was specially designed for her, from the printed torso and sloped skirt to her hair piece with a purple bandana and golden beads – not to mention her two tarot cards. Her serene expression suggests she is about to make a bold prediction...

MINI STATS

Theme
LEGO® Minifigures

Year
2013

First appearance
LEGO Minifigures
Series 9

Rarity

NOT ALL DOOM AND GLOOM

As well as the Tower card, which warns of impending disaster, the Fortune-Teller also carries the Sun card, which foretells good fortune.

For many years there were no minifigures with yellow hair, because it would have been the same colour as their faces. A lighter colour known as "cool yellow" was introduced in 2004 and has been used on a succession of blonde minifigures – but few wear it as timelessly as the Hollywood Starlet!

AND THE AWARD FOR...

The Hollywood Starlet is the fourth collectible Minifigure to come with a minifigure-shaped trophy, after the Karate Master, the Sumo Wrestler, and the Soccer Player.

MINI STATS

Theme
LEGO® Minifigures

Year
2013

First appearance
LEGO Minifigures
Series 9

Rarity

HOLLYWOOD STARLET

> I'M READY FOR MY CLOSE-UP NOW...

LOOKALIKE
The Hollywood Starlet could almost be mistaken for the legendary movie star Marilyn Monroe.

1 New blonde hair piece

2 Gold minifigure trophy

3 Glitzy strapless bodice and silver necklace

4 Slope piece represents floor-length gown

BLONDE ON YELLOW!

This muscular Minifigure delights circus crowds with his shows of superior strength, making his painted barbells look as light as LEGO feathers! His asymmetric leopard-print leotard shows off his sculpted physique, built up over many years in the circus. His moustache is also quite a spectacle, though it is in a style shared with several minifigures, including the Mariachi.

STRONG AND STYLISH

MINI STATS

Theme
LEGO® Minifigures

Year
2017

First appearance
LEGO Minifigures Series 17

Rarity

CIRCUS STRONGMAN

1 New face print with thick eyebrows and crow's feet

2 Belted leotard with a muscular chest underneath

3 Solid printing around lower half of legs looks like black boots

ICE-CREAM DREAM

Another welcome sight at any circus or fun fair is ice cream! The Ice-Cream Vendor in the People Pack – Fun Fair (set 60234) is dressed in colours that match her ice-cream flavours.

WINGS FOR ARMS!

YOU CRACK ME UP... AND THAT'S NO YOLK!

① Unique chicken head piece

② Minifigure face shows through mask

③ Moulded wings for arms first seen in 2013

WING TWINS

The same wings appear in pale yellow in 2019's Chicken Skater Pod (set 853958).

CHICKEN SUIT GUY

WILD ABOUT ANIMALS!

Chicken Suit Guy is the fourth animal costume collectible Minifigure, following Gorilla Suit Guy, Lizard Man, and Bunny Suit Guy.

The wacky Chicken Suit Guy's head piece – complete with beak, comb, and wattles – is not the only element that had never been seen before he was hatched. His wing-shaped arms were also new, but have since been used in a yellow hue on a skateboarding chicken!

MINI STATS

Theme
LEGO® Minifigures

Year
2013

First appearance
LEGO Minifigures Series 9

Rarity

ICE MATES
The Hockey Player (also in Series 4) wears the same blades as the Ice Skater.

1 Blonde hair piece with glamorous flick

2 Unique face print has stars around the eyes

3 Leotard print under skirt

4 Unique fabric skirt

ICE SKATER

With her winning smile and winning attitude – not to mention her specially designed blonde bun hair piece – the Ice Skater makes quite an impression. She has a distinctive fabric skirt, and her torso has one blue arm and one yellow arm, for a stylish off-the-shoulder look.

MINI STATS

Theme
LEGO® Minifigures

Year
2011

First appearance
LEGO Minifigures Series 4

Rarity

BUNS IN ABUNDANCE
The Ice Skater's hair is tied up in a tidy bun. Her hair mould was new in 2011, and has since been worn by the Fairy in brown, the Flamenco Dancer in black, and Wilma Flintstone in orange.

MINI STATS

Theme
LEGO® Minifigures

Year
2016

First appearance
LEGO Minifigures
Series 16

Rarity

A masterful Mexican musician, the Mariachi spreads love and happiness wherever he goes with his romantic rhythms! He has the privilege of playing a new acoustic guitar piece, which he can hold with both hands.

FIRST ACOUSTIC GUITAR

I WILL SERENADE YOU NOW.

MARIACHI

1. Black sombrero with pattern along the brim

2. Smart suit jacket with a red bow tie and silver belt buckle

3. Printed guitar piece has a peg to hold underneath its body

4. Silver pattern continues on sides of legs

STREET STYLE

In an orange scarf and check shirt, the Street Performer in the Capital City (set 60200) may not look as glamorous as the Mariachi, but he is also a talented musician.

113

Ladies and gentlemen, meet the LEGO weight champion of the world! In a dazzling gold championship belt to match his equally dazzling eye paint and spandex costume, the Wrestling Champion is celebrating a successful smackdown.

MINIFIGURE MULLET!

MINI STATS

Theme
LEGO® Minifigures

Year
2016

First appearance
LEGO Minifigures
Series 15

Rarity

WRESTLING CHAMPION

NO.1 WINNER
The Football Player in Minifigures Series 8 first lifted the trophy piece.

1 New mullet hairstyle is longer at the back

2 Flash-of-lightning eye paint and clenched teeth

3 Purple boots with lightning-flash sides

TRAINING PARTNERS

The LEGO® Creator set Downtown Diner (set 10260) includes a boxing gym, where this boxer and bodybuilder can be found working out.

BALLERINA

> MY DANCING IS ON POINT(E)!

1 Tidy bun printed with delicate white flowers

2 Ruffled white tutu piece fits on top of hips

3 Ribbons from silver pointe slippers wrap around legs

NEW TUTU

It is not easy for a minifigure with square feet to balance gracefully on her tiptoes! The Ballerina practises hard to maintain her perfect poise and posture. Her swan-like white outfit includes a frilly tutu piece for the first time!

JUST DANCE!

With bigger hair, brighter colours, and more spandex than the Ballerina, the Dance Instructor from Minifigures Series 17 looks very different, but she loves to dance just as much.

MINI STATS

Theme
LEGO® Minifigures

Year
2016

First appearance
LEGO Minifigures Series 15

Rarity

115

DANGER IS MY MIDDLE NAME!

1 Cool new hair piece

2 Devil-may-care face printing includes raised eyebrow and megawatt grin

3 "MF" on belt buckle stands for minifigure – or LEGO designer Michael Fuller

4 Helmet can be worn instead of hair piece

DAREDEVIL

HAIR OR HELMET?

Just two other collectible Minifigures have helmets and hair pieces: Intergalactic Girl and the Race Car Driver.

The Daredevil is all about style – check out his coiffured hair, handlebar moustache, and red, white, and blue jumpsuit. His head has two faces: a cockily confident one, and a far more appropriately worried expression...

MINI STATS

Theme
LEGO® Minifigures

Year
2012

First appearance
LEGO Minifigures Series 7

Rarity

MINI STATS

Theme
LEGO® NEXO KNIGHTS™

Year
2016

First appearance
Jestro's Evil Mobile (70316)

Rarity

Jestro may be a jester, but he is no laughing matter! He was the main villain in the LEGO NEXO KNIGHTS theme before later redeeming himself. This first variant is Jestro at his meanest, with skull-shaped bells around his neck, a ripped striped costume, and an unsettlingly large grin! He appears in just two 2016 LEGO sets.

UNFUNNY FUNNYMAN

JESTRO

HEY, WANT TO HEAR A JOKE?

CROOK BOOK

Jestro comes with the powerful Book of Monsters, which turned him into a villain.

① Jester's hat with a pom-pom at each end

② White face make-up and red lipstick

③ Tattered cape shared with fellow NEXO KNIGHTS villain General Magmar

TRICKSTER

The Jester from Minifigures Series 12 really is here for the laughs. Pay close attention to his card tricks on his unique tile playing cards!

117

WELL, I'M HAVING FUN!

CREEPY CLOWN!

① Top hat is new in magenta pink

② Neck ruffle first seen on the Actor from Minifigures Series 8

③ Scary green Gloombie hands

④ Gloombie hip extender makes Terry Top loom above other minifigures

TERRY TOP (POSSESSED)

One of two clowns in the LEGO Hidden Side Haunted Fairground set, Terry Top is recognizable by his bright-pink top hat. For once, the clashing colours of his costume aren't the most eye-catching thing about him – he has been possessed by a ghost!

CLOWN ETIQUETTE

Fellow clown Jimbo Limbo also haunts the Haunted Fairground set. He has just one Gloombie hand, so the other can hold an umbrella!

MINI STATS

Theme
LEGO® Hidden Side

Year
2020

First appearance
Haunted Fairground (70432)

Rarity

CHAPTER SEVEN
ONE OF A
KIND

SOME MINIFIGURES ARE JUST A LITTLE BIT DIFFERENT. THESE CHARACTERS REALLY STAND OUT FROM THE CROWD!

Released to celebrate 60 years of the LEGO® brick in 2018, Brick Suit Girl has a LEGO 2x3 brick body. With the exact dimensions of a regular 2x3 brick, she can attach to other pieces and also her friend in LEGO Minifigures Series 18 – Brick Suit Guy, who is dressed in the same brick suit in red.

BIRTHDAY SUIT!

STUD BUDDIES

Brick Suit Guy can literally stick by his friend Brick Suit Girl's side. He has an equally happy smile and tousled dark-brown hair.

BRICK SUIT GIRL

1 Side ponytail hair in dark brown for the first time

2 Wide, joyful smile – fitting for a LEGO celebration!

3 Brick torso with a neck stud and arms attached at sides

STICK-ON STUD

Brick Suit Girl has a red 1x1 plate as her accessory, which can also attach to her suit.

MINI STATS

Theme
LEGO® Minifigures

Year
2018

First appearance
LEGO Minifigures Series 18

Rarity

SAUSAGE ANDWICH BOARD!

1 Aptly named "sandwich board" piece shaped like a hot dog

I APPROACH MY JOB WITH RELISH!

1

2

2 Cheerful expression... Mmm, hot dogs!

3 Plain tan torso matches bun colour

3

"The initial design didn't have any mustard, but the final version has just the right amount of zig-zag sauce!"
GITTE THORSEN, LEGO DESIGN MASTER

HOT DOG MAN

This fast-food fanatic was the first collectible Minifigure to dress up as food! The detailed hot dog outfit was brand new in 2015. The sausage, bun, and mustard are all one piece that fits over his minifigure head.

SERVICE WITH A SMILE

Hot Dog Man isn't the only frankfurter fan in the collectible Minifigures theme. The Hot Dog Vendor is also sizzling with excitement about them. He cheerfully serves up a hot dog and soda in Series 17.

MINI STATS

Theme
LEGO® Minifigures

Year
2015

First appearance
LEGO Minifigures Series 13

Rarity

STRANGE SYMBOLS

Krakenskull's armour features symbols from an ancient time.

1 Skull-shaped helmet holds a printed 1x1 round tile

2 Mismatched eyes under helmet

3 New pointed cape design has seven strands of fabric

LORD KRAKENSKULL

MINI STATS

Theme
LEGO® NEXO KNIGHTS™

Year
2017

First appearance
Aaron's Rock Climber
(70355)

Rarity

A petrifying sight, Lord Krakenskull is the leader of an army of Stone Monsters in the world of LEGO NEXO KNIGHTS. His cracked stone armour and helmet are surging with electrical energy, which he uses to turn people to stone.

MOLTEN MONSTER

What's even scarier than a monster made from stone? A monster made from red-hot rocks and molten lava! General Magmar is the fiery leader of the Lava Monsters in the LEGO NEXO KNIGHTS theme. His army serve Jestro, a trouble-making jester, whose cackling image is on his belt.

Treasure hunters, pay attention!

You've hit the jackpot if you find a Mr Gold among your minifigures. Mr Gold has a gold face, gold torso, gold leg piece, and gold hat. The only parts that aren't glistening with a chrome gold finish are his white-gloved hands!

5,000 FOR 10
Created to mark the 10th series of collectible Minifigures, only 5,000 of Mr Gold were ever made. No one knows if all of them have been found. Do you have one in your collection?

GLOBE TROTTER
Anyone who found Mr Gold in 2013 was invited to enter their location on LEGO.com, to record where he turned up!

MR GOLD

I DARESAY IT'S YOUR LUCKY DAY!

1 No other minifigure has a gold top hat!

2 Unique, glad-to-be-gold expression

3 White hands can grasp his golden cane

MINI STATS

Theme
LEGO® Minifigures

Year
2013

First appearance
LEGO Minifigures Series 10

Rarity

CHROME SWEET CHROME

There's a new addition in the minifigure family! This skilled storyteller was the first minifigure to appear with a swaddled baby piece. The child can be held in the Tribal Woman's hand, or attached to a stud on her back to leave her hands free.

HEY BABY

MINI STATS

Theme
LEGO® Minifigures

Years
2016

First appearance
LEGO Minifigures
Series 15

Rarity

TRIBAL WOMAN

LEGO LIMBS
Dual-moulding allows for different colours on arms and legs.

MEET MY NEW ARRIVAL!

1 Coloured feathers slot into back of hair

2 Baby has simple face printing showing a happy smile

3 Dual-moulding allows boot colour to continue all the way around the leg

HAIR PAIR
The Tribal Hunter from LEGO Minifigures Series 1 wears the same hair piece as the Tribal Woman. It was first used in the LEGO® Western theme.

IT'S SANTA!

LEG-O-HO-HO!

SANTA CLAUS

1 Until 2012, all Santa minifigures wore this red hat, commonly seen on LEGO pirates!

2 Long white beard first worn by Majisto the Wizard from the LEGO® Castle theme

3 Black basket with neck bracket has space for a 1x2 LEGO tile

4 In the 2010 LEGO® City Advent Calendar, Santa has bare yellow legs, as he is taking a shower!

A SLEW OF SANTAS!
Including LEGO® *Star Wars*™ variants, there have been 25 different Santa minifigures!

Here's a minifigure you'll really want to find under your Christmas tree. The first in a long line of Santa minifigures, this limited-edition gift-giver comes with a sleigh laden with LEGO bricks, and looks just like the real thing!

ADDITIONAL CLAUS

The eighth series of collectible Minifigures included a new-look Santa Claus with a specially designed hat, sack, and printing. He also appears in Santa's Workshop (set 10245), but with a red sack.

MINI STATS

Theme
LEGO® Town

Year
1995

First appearance
Santa Claus and Sleigh (1807)

Rarity

WANT A PIZZA THE ACTION?

PIZZA FEATURES

1 New head print with stubble moustache and goatee

2 Printed pepperoni, black olives, and cheese

3 Checked trousers in the colours of the Italian flag

KNEAD A PIZZA?

He comes with a LEGO tile flyer to hand to hungry minifigures.

Who can resist a large slice of pizza? Pizza Costume Guy hopes no one can, as he earns his dough by advertising a pizza restaurant! His headgear was first seen as a pink slice of watermelon in THE LEGO® MOVIE 2™ line of collectible Minifigures in 2019, and later retopped as freshly baked pizza.

MINI STATS

Theme
LEGO® Minifigures

Year
2019

First appearance
LEGO Minifigures
Series 19

Rarity

PIZZA, PRONTO

Prefer your pizza delivered to your door? Pizza Delivery Man from LEGO Minifigures Series 12 comes with a printed-tile pizza box and a pepperoni pizza – let's hope it's still piping hot!

Theme
LEGO® NINJAGO®

Year
2014

First appearance
NinjaCopter (70724)

Rarity

Primary Interactive X-ternal Assistant Life-form, or P.I.X.A.L., was created as a robot aide for an antagonistic inventor in the world of LEGO NINJAGO, but soon lent her helping hands to the Ninja instead. Assembled in 2014, this variant of P.I.X.A.L. appears in just one set.

RARE ROBOT

P.I.X.A.L.

TWO HEADS ARE BETTER THAN ONE!

1 Swept-up ponytail hair piece seen in silver for the first time in 2014

2 Double-sided face print with red eyes on reverse

3 Electrical circuits visible on torso and face

A TOXIC ENEMY!

Toxikita, the sneaky villainess from LEGO® Ultra Agents, uses the same hair piece as P.I.X.A.L. but in green.

TAKE YOUR P.I.X.
In the LEGO NINJAGO TV show, inventor Cyrus Borg created 15 P.I.X.A.L. prototypes before this one, but they all went wrong!

This dapper chap is the leader of a brave band of Monster Fighters tasked with preventing a powerful vampire from wreaking havoc on the world. Rodney and his fellow fighters sport many brand-new printed elements, but Rodney goes one better, with a whole new leg entirely!

UNIQUE ROBOT LEG!

MINI **STATS**

Theme
LEGO® Monster Fighters

Year
2012

First appearance
The Vampyre Hearse (9464)

Rarity

DR RODNEY RATHBONE

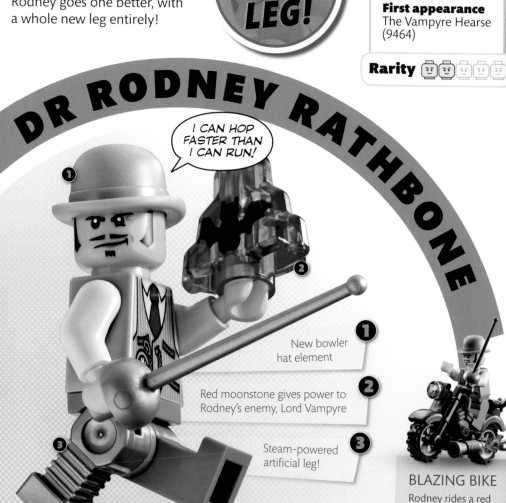

I CAN HOP FASTER THAN I CAN RUN!

1 New bowler hat element

2 Red moonstone gives power to Rodney's enemy, Lord Vampyre

3 Steam-powered artificial leg!

BLAZING BIKE

Rodney rides a red motorcycle while in hot pursuit of The Vampyre Hearse.

128

SQUARE **AND** RARE!

BEEP-BOOP! WHO-WANTS-A-HUG?

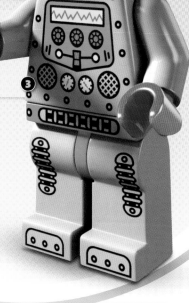

CLOCKWORK ROBOT

1 Rectangular head mould shared with Lady Robot from Minifigures Series 11

2 Wind-up key attaches to neck bracket

3 Printed details include torso rivets and toecaps

GET A-HEAD

Minifigures Series 6 introduced three new head moulds: for Clockwork Robot, Classic Alien, and Minotaur.

What makes a Minifigure tick? In this case, it's the key in his back! This rarely seen piece attaches to a neck bracket by means of a sideways stud. His rectangular head piece has a mouth full of sparking lights and glowing blue eyes that match his colourful torso print.

SILVER STREAK

The Clockwork Robot was preceded by the Robot from Minifigures Series 1, released in 2010. He has twin antennas and a special tool arm.

MINI STATS

Theme
LEGO® Minifigures

Year
2012

First appearance
LEGO Minifigures Series 6

Rarity

WEARABLE HORSE!

I AIN'T HORSIN' AROUND!

① New cowboy hat with upturned brim

② Fringed sleeveless shirt with black-and-white cowhide panels

③ Black horse hooves printed on his toes!

COWBOY COSTUME GUY

Well, howdy partner! This rootin' tootin' Minifigure rode straight out of the Wild West wearing a LEGO horse costume. Its painted head and front hooves hang over Cowboy Costume Guy's neck, and its long tail fits above his hips. Yeehaw!

ROUND 'EM UP

Cowboy Costume Guy wasn't the first cowboy around the campfire – several Wild West characters appear in the Minifigures line, including the lassoing Cowgirl from Series 8.

FLOWER POWER

Flowerpot Girl blossomed in LEGO Minifigures Series 18! Positioned in a ceramic pot skirt with a stalk-printed torso, six pink petals have bloomed on her specially designed headpiece. She is the second plant-themed Minifigure, after Series 14's Plant Monster.

MINI STATS

Theme
LEGO® Minifigures

Year
2018

First appearance
LEGO Minifigures
Series 18

Rarity

GROWING PAINS?
Flowerpot Girl has a nervous expression on the other side of her head.

FLOWERPOT GIRL

1 New head mould has green leaves around the back

2 Lime hands add more greenery!

3 Soil-coloured legs under new "pot" piece

FLORAL FRIENDS
Another green-thumbed garden fan starred in Minifigures Series 19. The Gardener comes with a leaf-stalk accessory and the first LEGO flamingo – though hers is a realistic-looking lawn ornament!

LEGO minifigures met Japanese manga in the exciting LEGO EXO-FORCE theme. Robot buster Hikaru has a new, angular hair piece made from rubber, and a dual-sided head showing two facial expressions (serious and angry). Both face prints show him wearing an orange visor.

MANGA-INSPIRED HERO!

MINI STATS

Theme
LEGO® EXO-FORCE™

Year
2006

First appearance
Hikaru's Training Glider (5966)

Rarity

HIKARU

DON'T TOUCH THE HAIR!

❶ Stylized spiky rubber hair

❷ Manga-style features printed on both sides of head

❸ Printed torso features the LEGO EXO-FORCE logo

COSTUME CHANGE

A variant Hikaru in a dark-blue suit featured in 2007's Sky Guardian (set 8103). He debuted a new white suit in 2008's Chameleon Hunter (set 8114).

POWER BRICKS

Every 2006 LEGO EXO-FORCE set featured a light-up brick called a power core.

> I PUT THE SPY IN SPIDER!

1 Spider legs attach to neck bracket

2 Frightening face print is part spider!

LEGO LEGEND

Spyclops takes his name from the Cyclops, a one-eyed creature from legend.

3 Just one arm is metallic silver

4 Pocket printing on mismatched legs

SPYCLOPS

Attention all Ultra Agents! This menacing minifigure is a mash-up of man, machine, and spider. Spyclops has fearsome fangs and two terrifying spider legs sprouting from his back!

MINI STATS

Theme
LEGO® Ultra Agents

Year
2015

First appearance
Spyclops Infiltration (70166)

Rarity

ALL LEGS

A similar character called Spy Clops appeared in the LEGO® Agents theme in 2008. He has four eyes and his torso sits on an awesome six-legged spider-leg assembly!

133

I'M THE ICING ON THE CAKE!

1 Pearl-gold party hat matches his bow tie

2 Blue suit jacket splattered with pink icing (his trousers inside the cake are almost all pink!)

3 Two-tier cake has a slot for Cake Guy to emerge from

CAKE GUY

MINI STATS

Theme
LEGO® Minifigures

Year
2018

First appearance
LEGO Minifigures Series 18

Rarity 😀😀😀😀😀

Is there any better way to celebrate 40 years of the modern LEGO minifigure than to see Cake Guy burst out of a big birthday cake? Wearing a dazzlingly wide smile, a jazzy party hat, and a snazzy party suit that is now covered in cake, this Minifigure is here to party!

READY TO ROCKET!

Every character in LEGO Minifigures Series 18 had a party theme – there were cakes, costumes, and entertainers. Firework Guy was there to ensure the celebrations went off with a bang!

Despite calling himself King of the Squidmen, this alien outlaw is in fact the one and only LEGO squidman! He first busted out in 2009, as part of the Space Police III subtheme, and his bulging upturned red eyes, fanged mouth, and long tongue make him a very collectible villain!

SCARY SQUID!

SQUIDMAN

NOT GILL-TY, YOUR HONOUR!

1 Unusual new head mould

2 Red cape also worn by Superman minifigure

3 Banknote accessory. Stop, thief!

"Squidman got me my job at the LEGO Group! I created him in a hiring workshop in London in 2005. The original design used a LEGO® Technic part as the head. Later, I was able to spend time perfecting it."
TIM AINLEY, LEGO CONCEPT MANAGER SPECIALIST

SQUID ON SCREEN

Squidman can be seen causing trouble in the LEGO DVD movie *The Adventures of Clutch Powers*.

135

The flute-playing, forest-dwelling Faun is a mythical creature like no other. With hooves, horns, and a hairy torso, half of him resembles a goat, while the other half is more minifigure-like. His flute is a regular bar piece with five holes printed on it.

GOAT LEGS

MINI STATS

Theme
LEGO® Minifigures

Year
2016

First appearance
LEGO Minifigures
Series 15

Rarity

FAUN

I'M TRYING OUT A GOATEE – WHAT DO YOU THINK?

FULLY FLEXIBLE
Many LEGO animals have static bodies, but the Faun's legs move back and forth like a regular minifigure's.

1 New hair piece with curly horns and protruding ears

2 Muscular torso with curly chest hair

3 Specially designed goat legs have fur and hooves

FURRY STORY
You might have heard the legend of Bigfoot, but what about Square Foot? This reclusive creature was spotted in Minifigures Series 14. His head piece is the same as the Yeti in Series 11 with brown, shaggy fur.

TWINKLE TOES AND TUSKS

AM I THE ELEPHANT IN THE ROOM?

ELEPHANT COSTUME GIRL

1 New elephant mask piece features short tusks and an upturned trunk

2 New mouse accessory with printed eyes and a tiny pink nose

3 Tiny white elephant toenails!

MINI STATS

Theme
LEGO® Minifigures

Year
2018

First appearance
LEGO Minifigures
Series 18

Rarity

It is a little-known fact that elephants can dance – and Elephant Costume Girl is out to prove it! In her pink tutu, also worn by Fairy Batman in THE LEGO® BATMAN MOVIE Minifigures Series 2, she is surprisingly light on her feet. Even her pet mouse is part of her performance!

RAINBOW BEAR

Another addition to the collective of animal-costumed characters in the Minifigures line, Bear Costume Guy from Series 19 has the same head mould as Panda Suit Guy, in a cacophony of colour!

I'M REALLY GOOD AT THE ROBOT DANCE.

PRICKLY CHARACTER!

① Long cactus body fits over her head and covers torso

② Head has an uncomfortable expression on the other side

③ Branching cactus arms, at awkward angles!

ARM ANOMALY
Minifigures with non-standard arms are rare – and minifigures without hands are even rarer!

CACTUS GIRL

MINI STATS

Theme
LEGO® Minifigures

Year
2018

First appearance
LEGO Minifigures Series 18

Rarity

With spikes all over her costume, Cactus Girl isn't the most popular person at a party – which is a shame because she could do with a hand! Her cactus arms are special new arm pieces that fit into her torso, which is printed with a green striped jumper under her ridged "sandwich-board" body.

BANANA SANDWICH

Cactus Girl is one of several minifigures who wear large sandwich-board costumes. Banana Suit Guy from Series 16 has a costume that "splits" over his torso!

BANANA!

CHAPTER EIGHT
WILD AT HEART

TAKE A WALK ON THE WILD SIDE WITH THESE AMAZING ANIMAL-THEMED CHARACTERS!

The General of the LEGO NINJAGO Anacondrai tribe, Pythor has a long Serpentine tail piece in place of legs. Both this chunky tail and his gaping head piece were created especially for him. He may well be the last remaining member of his tribe, but there are multiple variants of him, including two striking white versions with purple markings.

THE LAST OF HIS KIND

PYTHOR

JOIN MY FANG CLUB!

OPEN WIDE!
Pythor is the only Serpentine General to have a head with an open mouth.

1 Coiled head piece

2 Unique printing features gold scales and gemstones

3 Slithering tail piece made from a combination of plastic and rubber

HE'S ALL-WHITE!

This 2015 variant shows Pythor drained of colour because he had been swallowed by an ancient serpent! Another white variant appeared in 2016.

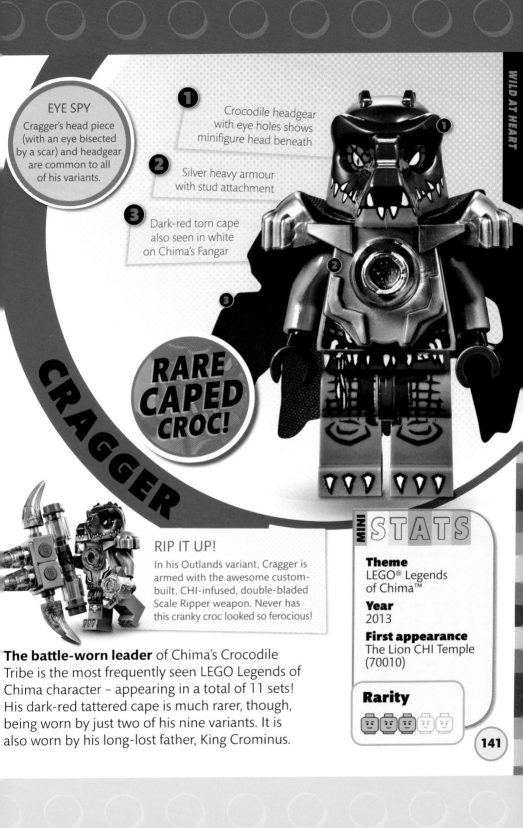

EYE SPY
Cragger's head piece (with an eye bisected by a scar) and headgear are common to all of his variants.

1 Crocodile headgear with eye holes shows minifigure head beneath

2 Silver heavy armour with stud attachment

3 Dark-red torn cape also seen in white on Chima's Fangar

RARE CAPED CROC!

CRAGGER

RIP IT UP!
In his Outlands variant, Cragger is armed with the awesome custom-built, CHI-infused, double-bladed Scale Ripper weapon. Never has this cranky croc looked so ferocious!

The battle-worn leader of Chima's Crocodile Tribe is the most frequently seen LEGO Legends of Chima character – appearing in a total of 11 sets! His dark-red tattered cape is much rarer, though, being worn by just two of his nine variants. It is also worn by his long-lost father, King Crominus.

MINI STATS

Theme
LEGO® Legends of Chima™

Year
2013

First appearance
The Lion CHI Temple (70010)

Rarity

141

STEP INTO MY PARLOUR...

SPINLYN

1 Red spider hidden inside abdomen

2 Headgear has six eyes, with holes for two on the head

3 Torso printed with golden spider image

4 Giant spider-like lower body adds six limbs

MINI STATS

Theme
LEGO® Legends of Chima™

Year
2014

First appearance
Spinlyn's Cavern (70133)

Rarity

The self-styled Queen of the Spider Tribe, Spinlyn is the only Chima minifigure not to have standard legs. The six legs of her spider body and the two arms on her torso give her the necessary total of eight limbs, and make her one of the biggest minifigures of all.

"Using the meteorite piece as her abdomen works really well. You can put baby spiders inside!"
ALEXANDRE BOUDON, LEGO DESIGN MASTER

This Yeti's new head mould covers most of his torso. Every one of his elements features light royal-blue (except for his transparent-blue popsicle), making him the second Minifigure to use this colour – the other being the Scientist from Minifigures Series 11.

KEEPS A COOL HEAD!

MINI STATS

Theme
LEGO® Minifigures

Year
2013

First appearance
LEGO Minifigures
Series 11

Rarity

CLEVER CREATURE

The Yeti is one of the Master Builders seen in THE LEGO® MOVIE™.

YETI

1 New head mould, since used for Breezor the Beaver in LEGO Legends of Chima

2 Popsicle in this shade of blue is unique to the Yeti

3 Light-blue printing for shaggy fur effect

ICE TO MEET YOU

The first winter-themed character in the Minifigures theme was the Skier from Series 2. He's easily identifiable out on the slopes – his blue ski suit with a numbered bib is unique to him.

This fierce feline appeared in the monsters-themed LEGO Minifigures Series 14 – perhaps a sign that she can be pretty fearsome if her tail-like whip comes out! With pointy ears in her hair and stripes from her cheeks to the tip of her tail, she is one of a pack of animal-costumed characters that started with Gorilla Suit Guy.

TONS OF TIGER STRIPES

MINI STATS

Theme
LEGO® Minifigures

Year
2015

First appearance
LEGO Minifigures Series 14

Rarity

TIGERWOMAN

MY FAVOURITE SONG? I'D HAVE TO SAY "EYE OF THE TIGER".

1 New hair piece with feline ears and flicked-out ends

2 Flexible whip accessory

3 Furry white chest and cinched-in waist belt

4 Tiger tail fits between hips and torso

TIGER FACE

A little girl from a LEGO® City Fun Fair (set 60234) has stripy tiger face paint!

144

TRANSPARENT BLUE BODY PARTS

MAULA

1 Battle-scarred, grey-coloured mammoth head piece with large white tusks

2 Transparent-blue armoured chestplate

3 Detailed torso print visible beneath transparent piece

4 One transparent-blue leg and one grey leg

"I spent a lot of time on Maula's braids and the ornaments on her torso."
TORE HARMARK-ALEXANDERSEN, LEGO DESIGNER

MINI STATS

Theme
LEGO® Legends of Chima™

Year
2014

First appearance
Maula's Ice Mammoth Stomper (70145)

Rarity

This force of nature leads the Mammoth Tribe, and her minifigure looks just as formidable as you'd expect! Maula is unusual because some of her parts are transparent blue, making them look as if they have been carved from ice.

FROZEN FORTRESS

As well as commanding the Ice Mammoth Stomper, Maula also leads her tribe in their impressive but icy base, Mammoth's Frozen Stronghold (set 70226).

I'M A WOLF IN WOLF'S CLOTHING!

SHAPE-SHIFTER

LEGEND

Akita bears a resemblance to a *Kitsune* – a magical, fox-like creature in Japanese folktales.

1 Wolf mask and black hair braids are all one piece

2 Head features red whiskers and a fanged, wolf-like grin

3 Split cloth cape looks like long, red-tipped tails

AKITA

Theme
LEGO® NINJAGO®

Year
2019

First appearance
Castle of the Forsaken Emperor (70678)

Rarity

This shape-shifting formling from a snowy realm of LEGO NINJAGO can take the form of both a minifigure and a wolf – though not necessarily in the same set! Minifigure Akita has a specially designed cape representing her wolf form's tails.

LOYAL WOLF

Akita's wolf form is unmistakable, with matching whiskers and three tails. It first appeared in battle alongside Green Ninja Lloyd in 2019's Lloyd's Journey (set 70671).

MINI STATS

Theme
LEGO® Minifigures

Years
2019

First appearance
LEGO Minifigures
Series 19

Rarity

This mischievous monkey god leapt straight out of an ancient myth. The legendary Monkey King has supernatural strength, strong golden armour, and a staff that can extend during battle. He also has a very short temper – so watch out!

MONKEYING AROUND

The cackling Monkey King is a real trickster who loves to laugh at his own jokes!

MONKEY KING

MAGIC MONKEY

1 Red ribbon fits into a hole on the swept-back hair piece

2 Extendable staff features two LEGO® *Star Wars*™ lightsaber hilts

3 Cape has two stretchy strands that billow as he flies through the air

PRIMATE PAL

Before the Monkey King came along, Gorilla Suit Guy was the only primate animal in the LEGO Minifigures pack. He may look convincing, but he's really just a regular guy in a costume, holding a banana!

147

There was a buzz when this alien bug winged its way into the LEGO Space subtheme Galaxy Squad! The Mosquitoid's head is made from a special new mould, and this was the first time its style of wings had been seen on a minifigure, too.

SWARM FORCE

As well as Mosquitoids, Galaxy Squad also has to contend with Alien Buggoids (both the red and green kind). Where's a giant flyswatter when you need one?

MINI STATS

Theme
LEGO® Space

Year
2013

First appearance
Swarm Interceptor (70701)

Rarity

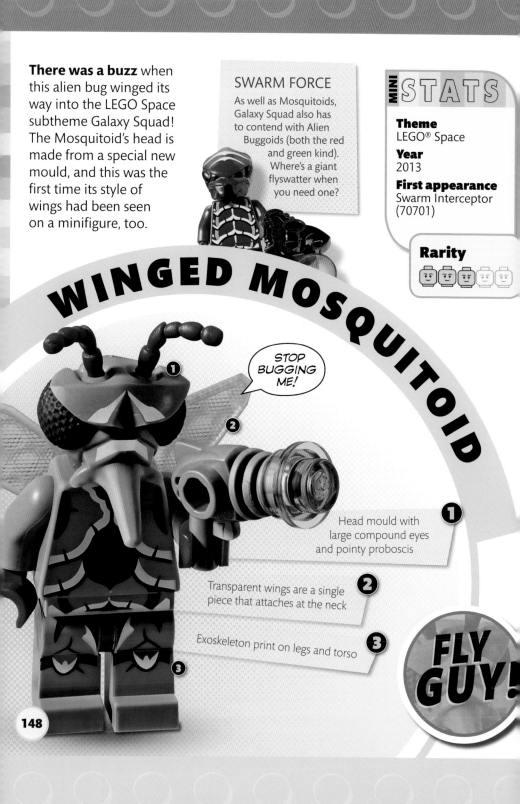

WINGED MOSQUITOID

STOP BUGGING ME!

1 Head mould with large compound eyes and pointy proboscis

2 Transparent wings are a single piece that attaches at the neck

3 Exoskeleton print on legs and torso

FLY GUY!

TWO
HEADS
ARE BETTER
THAN
ONE!

YESSS!

FANGPYRESSS FOREVER!

1 Rare two-headed element

2 White scale pattern runs from head to legs

3 Fangdam is the only Fangpyre with red legs

FANGDAM

Fangdam is the second-in-command of the Fangpyre tribe in the LEGO NINJAGO theme. A Fangpyre's bite can create a new serpent, and Fangdam's second head is the result of a bite from a fellow Fangpyre who mistook him for a slug!

FAMILY RESEMBLANCE

Fangdam's brother Fangtom has a near identical head piece, with one exception – Fangtom's has white markings on the back, rather than black. His sneaky sibling also has a long tail piece instead of legs.

MINI **STATS**

Theme
LEGO® NINJAGO®

Year
2012

First appearance
Fangpyre Truck Ambush (9445)

Rarity

149

1 Gorilla head over standard minifigure head mould

2 Standard minifigure hands hold giant fist weapons

GIANT METAL FISTS!

Fists also seen in the Ultra Agents and NINJAGO themes

G'LOONA

MINI STATS

Theme
LEGO® Legends of Chima™

Year
2013

First appearance
Gorzan's Gorilla Striker (70008)

Rarity 😀😀😀😐😐

LONE WOLF

Like G'Loona, Windra is the only female in her tribe to be made into a minifigure. She appears in just one set to date: Worriz's Combat Lair (set 70009).

G'Loona is the only known female minifigure of Chima's Gorilla Tribe, but what makes her even more special is her huge metal hand armour! She also has short, unposable legs and a striking costume decorated with braided leaves and pink flowers.

MINI STATS

Theme
LEGO® Minifigures

Year
2013

First appearance
LEGO Minifigures
Series 10

Rarity

This bee scores an "A" in the collectibility stakes, with unique headgear, special clear wings, and brand-new bumblebee stripes! She was also the very first female Minifigure to dress as an animal. Though she lacks a stinger, she has no shortage of honey in her printed pot.

CHOICE OF THE BEEHIVE!

BUMBLEBEE GIRL

A BUSY BEE'S WORK IS NEVER DONE!

1 New cap with insect antennae

2 Plastic wings attach to neck

3 Pot element also carried by Leprechaun Minifigure

WINGING IT
Adding a touch of magic to the Series 8 Minifigures, the Fairy wears her wings in transparent blue.

By the pleased look on his face, Shark Suit Guy knows his costume is fin-tastic – even if it is also pretty heavy! The other side of his head has an exasperated expression for when he gets tired. The shark body hangs from his jaw-some head piece all the way down the back of his body.

TOP TEN
Shark Suit Guy is the tenth LEGO minifigure to wear an animal costume.

FLAPPING ON ICE

Penguin Suit Guy appeared in Minifigures Series 16. He has the same arms as Shark Suit Guy, in black. They may help him balance on his ice skates!

SHARK SUIT GUY

1 Etched gills on either side of shark body

2 Fierce shark-jaw face hole!

3 Blue fin arms fit into regular arm holes in torso

4 Plain blue hips and legs blend in with shark tail

MINI STATS

Theme
LEGO® Minifigures

Year
2016

First appearance
LEGO Minifigures Series 14

Rarity

JAW-DROPPING COSTUME

152

HOT WINGS!

1 Blazing orange wings

2 Fiery red feathered head piece

3 Gold collar and wing harness, with flame icon

4 Slope piece features belt and pendants

FLUMINOX

PRETTY FLY...
Fluminox's gold shoulder armour was later worn by 2016's Flying Warrior Minifigure.

This Phoenix from the LEGO Legends of Chima theme wears an outfit befitting his role as leader of his tribe. He wears new shoulder armour with flame-coloured wings attached, and has a floor-length robe printed with ornate golden details.

LEG-ENDS OF CHIMA
Fluminox has a variant with legs instead of a slope piece, so he can sit on his Speedor bike in Inferno Pit (set 70155).

MINI STATS

Theme
LEGO® Legends of Chima™

Year
2014

First appearance
Flying Phoenix Fire Temple (70146)

Rarity

WILDLIFE PHOTOGRAPHER

1 Snow goggles printed on head

2 Cosy fur-lined hood was new for 2016

3 Red-and-blue snowsuit with multiple pockets

PENGUIN PIECES
The new penguin mould has a stud on its back so other pieces can be added to it.

The world is one big photo shoot for the globe-trotting Wildlife Photographer. Dressed for arctic conditions with her camera at the ready, she is prepared to wait for hours in the snow and ice for the perfect shot of a new colony of LEGO penguins – one of which first appeared alongside her!

TREE POSE
It takes patience and creativity to be a nature photographer. This guy from the Safari Off-Roader (set 60267) clearly has both – he is disguised as a tree!

MINI STATS

Theme
LEGO® Minifigures

Year
2013

First appearance
LEGO Minifigures
Series 10

Rarity

A monster straight out of classical myth, Medusa's hair is alive with snakes, and she even has a slithering tail. Her green-featured, fanged face has an angry-looking expression on the reverse of her head.

SNAKES FOR HAIR!

LEGEND!
Medusa is part of a mythical race of beings called Gorgons.

MEDUSA

LOOK INTO MY EYES...

1 New hair piece later seen in dark red for General Machia in LEGO NINJAGO sets

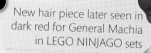

2 Entwined snakes trim the top of her fang-shaped tube top

3 Green tail piece has flexible rubber tip

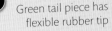

TWIST IN THE TAIL

Snake tail pieces were first used for the Serpentine General minifigures in LEGO NINJAGO sets.

Introduced in 2010, LEGO Atlantis featured many innovative new minifigure designs for its fierce underwater warriors. A brave Salvage Crew comes face to face with all kinds of fishy foes, but none is so memorable as the Squid Warrior and his tentacles of terror!

TERRIFIC TENTACLES

The terrifying Alien Commander in 2011's LEGO® Alien Conquest sets also uses the tentacle piece.

MINI STATS

Theme
LEGO® Atlantis

Year
2010

First appearance
Gateway of the Squid (8061)

Rarity

SQUID WARRIOR

SOMETHING SEEMS FISHY TO ME!

1 Elaborate squid head with printed yellow patterns

2 Bloodshot eyes printed on head beneath squid piece

3 Gold trident carried by all the underwater warrior minifigures

4 Large tentacle piece in place of standard legs

DEMONS OF THE DEEP

Other unusual undersea creatures encountered by the LEGO Atlantis Salvage Crew include Shark Warriors and Manta Warriors!

LOTS OF LEGS!

GOLDEN DRAGON MASTER

TEENAGE TEACHER

1 Samurai helmet, new in pearl gold in 2018

2 Elaborate dragon-shaped chestplate

3 Tied-on knee armour and sandals on legs

1

2

3

DRAGON MASTER

Whoever wears the Dragon Armour can control the mighty Firstbourne dragon.

Under the
Golden Dragon Master's helmet is a familiar face, albeit a younger version than usual. This is the teenage Master Wu, wise teacher of Ninja. He wears special armour that was created by his father, the first Spinjitzu Master.

RAWR!

Dragon Suit Guy in LEGO Minifigures Series 18 might be in a less powerful dragon suit than Master Wu, but he makes up for it with his roaring face!

MINI STATS

Theme
LEGO® NINJAGO®

Year
2018

First appearance
Dragon Pit (70655)

Rarity

MORE THAN A HORSE

GET TO THE POINT!

1 Pearl-gold unicorn horn shared with Astro Kitty from THE LEGO® MOVIE™ sets

2 Blue unicorn mask has a darker blue mane

3 Shield features a rearing unicorn – or is it dancing?

HAIRY TAIL

His dark-blue horse tail is worn by Unicorn Girl in lavender and Cowboy Costume Guy in brown.

UNICORN GUY

Unicorn Guy is not only a magical creature with a shimmering golden head horn, he is also a knight of a magical realm, with a shimmering golden sword! By the look on his winking face, he doesn't take his epic quests all that seriously.

MINI STATS

Theme
LEGO® Minifigures

Year
2018

First appearance
LEGO Minifigures
Series 18

Rarity

FAIRYTALE FIRST

Many people don't believe in unicorns, but Unicorn Girl was the first to silence the neigh-sayers! With a white horn and lavender mane, she appeared in Minifigures Series 13.

158

Theme
LEGO® Minifigures

Year
2019

First appearance
LEGO Minifigures
Series 19

Rarity

Fox Costume Girl is licking her lips, so perhaps it's time for her dinner. Is that it in her bulging sack, or will it be the LEGO chicken she appears with? Let's hope not – it's the first time the tan-coloured breed has been seen in a LEGO set!

FURRY FOX TAIL!

FOX COSTUME GIRL

THIS BAG? IT'S LIGHT AS A CHICKEN FEATHER!

 1 New fox mask with black ears and white cheek fur

 2 Orange bushy tail with a white tip is made from rubber

3 Black hands and feet look like fox paws!

EVEN NUMBERS

LEGO Minifigures Series 19 was the first to have an equal split of male and female minifigures.

BUSHY BUDDY

Wolf Guy from LEGO Minifigures Series 14 shares the same tail mould in dark grey.

159

This Minifigure thinks butterflies, bugs, and blossoms are fluterly wonderful. She wears specially designed transparent butterfly wings, cute pink flowers in her hair, and a tank top that features a smiling butterfly – who looks just as adorable as Butterfly Girl does!

LITTLE LEGS
Butterfly Girl has short legs in bright pink. The first minifigure ever to have these short legs was Yoda in 2002 LEGO® *Star Wars™* sets.

MINI STATS

Theme
LEGO® Minifigures

Year
2017

First appearance
LEGO Minifigures Series 17

Rarity

BUTTERFLY GIRL

MOTHER NATURE SCENT ME.

1 New blonde ponytail piece with printed pink flowers

2 Butterfly wings fit around her neck

3 Lime-green flower stalk was new for 2017

NATURE FAN
Before Butterfly Girl fluttered along, Spider Lady wove a web in Series 14 of the Minifigures theme, clutching her pet spider, Baron von Skitters.

BEAUTIFUL BUTTERFLY WINGS

160

CHAPTER NINE

ROTTEN ROGUES

EVEN SOME MINIFIGURES HAVE A DARK SIDE, AND THE ONES IN THIS ROGUES' GALLERY ARE BAD TO THE BASEPLATE!

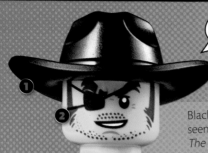

DON'T RATTLE ME!

1 Black cowboy hat first seen in 2013 in LEGO® *The Lone Ranger*™ sets

2 New face print with eyepatch, stubble, and sideburns

3 Claw necklace

SNAKE SYMBOL
Snake Rattler's jacket has a snake coiled in the shape of a dollar symbol on the back!

SNAKE RATTLER

This master crook is a wanted man in LEGO City. Look out for his eyepatch, cowboy hat, and distinctive jacket. He first attracted police attention in a Police Helicopter Chase (set 60243), where he escaped with a safe full of diamonds!

MINI STATS

Theme
LEGO® City

Year
2020

First appearance
Police Helicopter Chase (60243)

Rarity

HERO IN A HELMET
Sam Grizzled is responsible for wrangling the Snake Rattler. He has spent many years catching crooks for the LEGO City police – he won't let this one slither away!

Theme
LEGO® Agents

Year
2008

First appearance
Volcano Base (8637)

Rarity

No one knows quite why LEGO Agents villain Claw-Dette replaced her right arm with a claw, but it doesn't seem to have improved her temperament. The arm is made up of two pieces: a special silver arm piece – created for LEGO Agents – and a grey robot claw hand.

SHE'S GOT A CLAW FOR A HAND!

GET A GRIP!

1 Scary robot claw piece

2 Same face as Evil Witch from LEGO® Castle theme

3 No-nonsense bobbed hairdo with fringe

4 Zipped-up jacket with Dr Inferno's logo

SHARE THE HAIR
Nya from LEGO® NINJAGO® wears the same bob hair piece.

GIMME FIVE
Dr Inferno, Claw-Dette's boss, also has one robotic claw arm. They must do one hard-hitting high five when they celebrate!

163

This **LEGO NINJAGO** minifigure is actually two characters in one! On one side of his head piece, he is smiling Cyrus Borg, a skilled inventor. On the other side, he is transformed by technology into the villain OverBorg.

MINI STATS

Theme
LEGO® NINJAGO®

Year
2014

First appearance
OverBorg Attack
(70722)

Rarity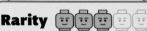

CYRUS BORG

ALLOW ME TO INTRODUCE MY SELVES...

1 Headgear features a robotic red rangefinder

2 Metallic plates patch up the left side of Borg's scowling face

3 Intricate wire, plate, and pipe printing on torso piece

4 One silver arm and one black arm reflect Borg's two characters

LESS-SINISTER CYBORG

Not all cyborgs are supervillains – the Cyborg from Minifigures Series 16 is a cool-headed hero. Like Cyrus, the left side of her head is robotic.

MYSTERY MAN

YOUR MONEY OR YOUR LEGO BRICKS!

1 Black tricorn hat, often worn by pirate minifigures

2 Black bandana masks a gruff expression!

3 Victorian-era black cloak with large collar

4 Flintlock pistol – new in dark grey

HIGHWAYMAN

Characteristically secretive, the masked Highwayman was a mystery figure in LEGO Minifigures Series 17. His identity was intriguingly blanked out in promotional materials for the series, and only revealed when fans unmasked him.

BALD-HEADED BUCCANEER

Perhaps the Highwayman stole his tricorn hat from Scallywag Pirate? He introduced a new swashbuckling style in Minifigures Series 16 – a bald head and bandana head piece!

MINI STATS

Theme
LEGO® Minifigures

Year
2017

First appearance
LEGO Minifigures Series 17

Rarity

I'VE GOT A PAIR OF SABERS!

1 Fangs match the colours of the head piece beneath

2 Transparent blue arm with opaque blue hand

3 Legs printed with belt, tokens, and armour detailing

4 Transparent-blue leg with printed claws

SIR FANGAR

CLEAR AS ICE

Transparent light-blue body parts were created especially for Legends of Chima.

The leader of the Saber-Tooth Tiger Tribe, Fangar is a truly fearsome-looking minifigure. He has battle scars on his face and long, mismatched teeth – but it's his icy body parts and armour that would really make you shudder!

MINI STATS

Theme
LEGO® Legends of Chima™

Year
2014

First appearance
Fire vs Ice (70156)

Rarity

TATTERED TWIN

A black version of Sir Fangar's cape is worn by another chilling villain: the Blizzard Sword Master minifigure in LEGO NINJAGO sets. He also has a matching icy arm.

MINI STATS

Theme
LEGO® NINJAGO®

Year
2018

First appearance
Temple of
Resurrection (70643)

Rarity

A member of the Ninjago royal family, Princess Harumi is dressed in a striking floral cape and kabuki face make-up. Her first LEGO appearance was in these robes, then she had a dramatic change – because Harumi, as the Ninja would discover, has a secret identity.

SECRETIVE PRINCESS

HARUMI

1 Face-framing gold tiara is part of hair piece

2 High-collared cape printed with red lotus flowers

3 Long green robes and gold zori sandals

THE QUIET ONE

Princess Harumi's secret identity first appeared in two 2018 sets. She is the "Quiet One", who leads the Sons of Garmadon – a villainous biker gang.

This feared forest dweller is part of a band of bad guys who roam LEGO woodland in search of riches. His brown-and-green clothing helps him blend in with the trees, and his hooded cowl hides his face – possibly because he has a not-so-scary smiling face underneath it!

WOLFPACK MEMBER?

Robbers wearing wolf symbols caused trouble in LEGO Castle sets from 1992. Could Rogue be one of them?

Theme
LEGO® Minifigures

Year
2016

First appearance
LEGO Minifigures Series 16

Rarity

ROGUE

HOODED WOODMAN

I KEEP SNAGGING MY TIGHTS ON THE TREES!

1 Quiver of arrows

2 Brown shawl secured with wolf-head pin

3 Gloved hand on bow-and-arrow firing arm!

4 Leather belt and satchel printed on legs

SECURITY!

A more modern thief but just as sticky-fingered, the Jewel Thief was seeking diamonds and other gems in Minifigures Series 15.

ENEMY OF THE AGENTS!

DR INFERNO

I'M HAIR TO CAUSE TROUBLE!

1 Hair like flames

2 Robot claw later worn by henchwoman, Claw-Dette

3 Scarred head piece shows excited grin

4 Stylish suit with flaming skull logo

NO ARM DONE

Before 2008, there had been minifigures with hook-hands, but none had alternative arm pieces.

MINI STATS

Theme
LEGO® Agents

Year
2008

First appearance
Mobile Command Centre (8635)

Rarity

With a striking new hair piece and a robotic claw, this grinning genius on the LEGO Agents' most-wanted list soon joined LEGO fans' most-wanted lists, too. But Dr Inferno is most notable for being the first LEGO minifigure with a non-standard arm.

A KNIGHT'S TALE

The very first minifigure villain charged into action 30 years before Dr Inferno. The Black Cavalry Knight was included in Castle (set 375). His smiling face is hidden by a black visor.

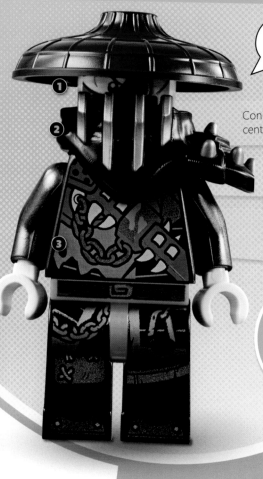

NEVER APPROACH ME WITH A MAGNET!

① Conical hat with raised centre first seen in 2017

② Spiky shoulder armour with metallic face bars

③ Rusting, spiked belt and chains across tattered clothing

HAT HAIR

A different variant of Heavy Metal has black hair in a bun instead of a hat.

HEAVY METAL

Part of a group of LEGO NINJAGO Dragon Hunters who salvage machinery, Heavy Metal (also known as Faith) wears some of her best finds. First seen in an identity-concealing conical hat, she later softens her exterior and befriends the Ninja.

MINI STATS

Theme
LEGO® NINJAGO®

Year
2018–2019

First appearance
Firstbourne (70653)

Rarity

IRON BARON

Heavy Metal is second-in-command to the Dragon Hunters' boss, the Iron Baron. He and his metallic peg leg first appeared aboard the Dieselnaut (set 70654) in 2018.

MINI STATS

Theme
LEGO® NEXO KNIGHTS™

Year
2017

First appearance
The Stone Colossus of
Ultimate Destruction
(70356)

Rarity

General Garg is a sinister stone statue brought to life by dark forces in the LEGO NEXO KNIGHTS theme. He has two jagged black wings with transparent blue "membranes". They fit into studs on the back of his shoulder armour.

GRUESOME
GARGOYLE

GENERAL GARG

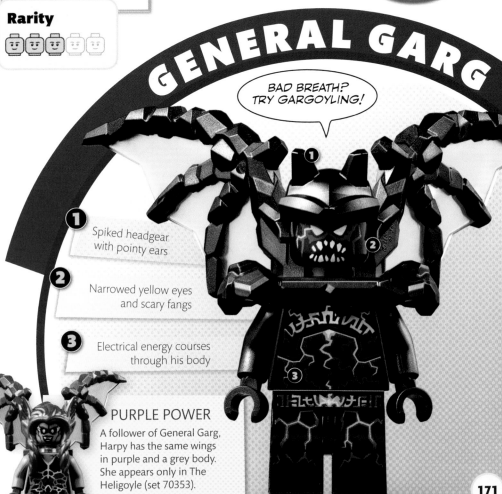

> BAD BREATH?
> TRY GARGOYLING!

1 Spiked headgear with pointy ears

2 Narrowed yellow eyes and scary fangs

3 Electrical energy courses through his body

PURPLE POWER

A follower of General Garg, Harpy has the same wings in purple and a grey body. She appears only in The Heligoyle (set 70353).

171

This menacing minifigure is the ruler of a digital world inside the Ninjago video game *Prime Empire*. Unagami wears an all-black robe with the appearance of a lit-up computer circuit board. His double-sided face is smiling on one side, and filled with pixelated rage on the other!

NEXT-LEVEL VILLAIN

MINI STATS

Theme
LEGO® NINJAGO®

Years
2020

First appearance
Empire Temple of Madness (6860)

Rarity

UNAGAMI

LIFE'S A GAME. LET'S PLAY!

1 Pulled-up ponytail with red hair wrap

2 Red dots and narrowed eyes on his angered expression

3 New pointed beard fits around his neck

4 Two-tiered shoulder armour in red for the first time

SKIRT FIRST

Unagami was the first NINJAGO minifigure to wear a new skirt piece, introduced in 2018 to replace the classic LEGO slope.

ROBOT ENFORCERS

Unagami has an army of mindless robotic Red Visors to do his bidding inside *Prime Empire*. Their distinctive transparent visors fit over a black helmet with a hole in the top.

ROCK-SOLID CREW

ROOG

1 Jagged head-and-shoulder armour is one piece, first seen on Roog and his brother

2 Facial features formed from cracks in the stone

3 Armoured fists in dark purple for the first time

ROLLING STONE
The eldest of the three brothers, Rumble, is more rock and roll than the other two – he is a monster truck with chomping jaws!

LITTLE BROTHER
They may have been cut from the same rock, but Roog and his youngest brother have quite different looks. Reex has plain, shorter legs, a black eyepatch over one eye, and different printing on his head-and-shoulder armour.

Meet the middle brother of a trio of stone-faced siblings. Roog, along with his brothers Rumble and Reex, petrifies the LEGO NEXO KNIGHTS' kingdom as part of the Stone Army. They are all exclusive to just one 2017 set. In it, the brothers battle Axl, who must be one very brave knight!

MINI **STATS**

Theme
LEGO® NEXO KNIGHTS™

Year
2017

First appearance
The Three Brothers (70350)

Rarity

CROSSING ME WOULD BE A MISSSSS-TAKE.

1 Transparent details give her hood and eyes a burning glow!

2 Double-headed cobra shoulder armour in pearl gold

3 Ghost body mould first used for the ghosts of Ninjago in 2015

GHOST SNAKE!

ASPHEERA

SNAKE ASSISTANT

As his name suggests, Char looks like he may have been near Aspheera's fire power for too long! A Pyro Viper, Char is Aspheera's assistant. He wears a copper version of his boss's shoulder armour.

This snake sorceress has the power to bring an old Ninjago foe, the Serpentine, back to life! Her flaming cobra hood and ghostly lower body displays the Fire elemental power she possesses. Aspheera first appeared with two fellow Pyro Vipers in the Fire Fang set.

MINI STATS

Theme
LEGO® City

Year
2013

First appearance
LEGO City Starter Set (60023)

Rarity

LEGO City has seen its fair share of bad guys over the years, but even its biggest villains have a lovable look! This classic Crook has a stylish moustache and wears the (not-so-stylish) striped shirt of an escaped prisoner.

STOP, THIEF!

CROOK

WHO WAS THAT MASKED MAN?

Eleven other minifigures have this masked face print, which must make police ID line-ups very difficult!

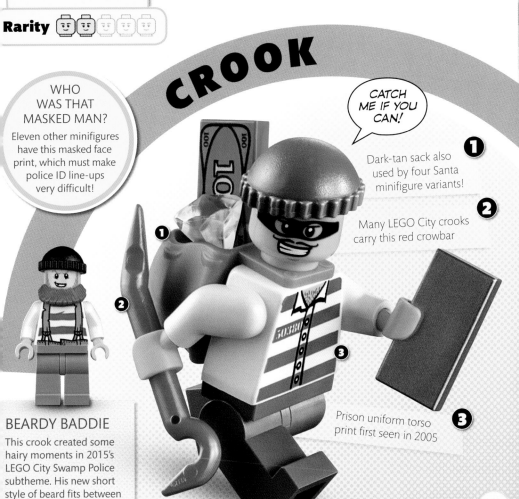

CATCH ME IF YOU CAN!

1 Dark-tan sack also used by four Santa minifigure variants!

2 Many LEGO City crooks carry this red crowbar

3 Prison uniform torso print first seen in 2005

BEARDY BADDIE

This crook created some hairy moments in 2015's LEGO City Swamp Police subtheme. His new short style of beard fits between his head and torso pieces.

175

Decked out all in black, the original Blacktron Spaceman is light years ahead of the classic LEGO astronauts that preceded him. Don't let his smile fool you, though – these stylish rocketeers were the first LEGOLAND Space bad guys, travelling the galaxy purely for profit.

FIRST SPACE VILLAIN

MINI STATS

Theme
LEGOLAND® Space
Years
1987–1991, 2009
First appearance
Invader (6894)

Rarity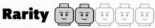

BLACKTRON SPACEMAN

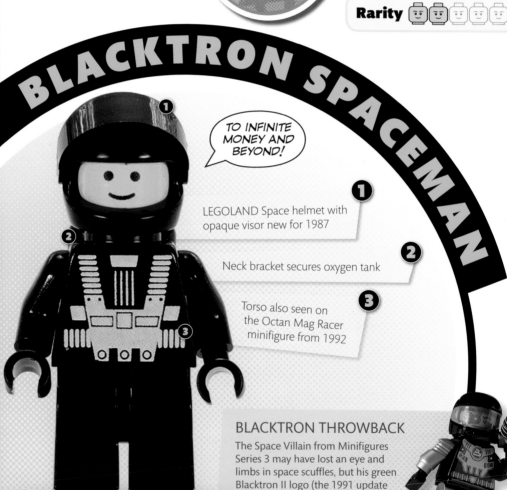

TO INFINITE MONEY AND BEYOND!

① LEGOLAND Space helmet with opaque visor new for 1987

② Neck bracket secures oxygen tank

③ Torso also seen on the Octan Mag Racer minifigure from 1992

BLACKTRON THROWBACK

The Space Villain from Minifigures Series 3 may have lost an eye and limbs in space scuffles, but his green Blacktron II logo (the 1991 update of the theme) remains unscathed!

176

MINI STATS

Theme
LEGO® NEXO KNIGHTS™

Year
2017

First appearance
Ruina's Lock & Roller (70349)

Rarity

1 Face-crackingly angry expression!

2 Electrified gown features mysterious symbols

3 Lower half of gown is a printed slope piece

MUM?
In the NEXO KNIGHTS TV series, Ruina is confirmed to be hero Clay's mother!

RUINA

This villain will really make your hair stand on end! A witch with the Stone Army in the LEGO NEXO KNIGHTS theme, Ruina has hair that is aglow with shocking blue electricity. The distinctive hair piece, with curls that tumble over one side of Ruina's face, was first seen in black in 2010, on Bellatrix Lestrange from the LEGO® *Harry Potter*™ theme.

HAIR-RAISING WITCH

STONE CLAY

This grey variant of knight Clay appears in The Stone Colossus of Ultimate Destruction (set 70356). He is turned to stone in battle and briefly fights for the Stone Army.

I'M STEALING GOLD TO PAY FOR A SHAVE!

HAT'S AMAZING!

All three hats in the Black Seas Barracuda set were brand-new designs – including one worn by the ship's figurehead!

① Three-cornered or "tricorn" hat

② Printed eyepatch instead of an eye

③ Printed stubble and shaggy hair

④ Classic pirate jacket, striped shirt, and belt

PIRATE

MINI STATS

Theme
LEGO® Pirates

Years
1989, 2002

First appearance
Black Seas Barracuda (6285)

Rarity

This pirate has the recognizable cheery smile of a classic minifigure, but also the scruffy hair, eyepatch, and stubble of a scoundrel! Added details like these were all part of a new kind of minifigure, introduced in the LEGO Pirates theme in 1989 – minifigures would never be the same again!

IS HAT A FACT?

LEGO headgear has been around even longer than the modern minifigure. Early LEGO figures, like this one released in 1975, wore a hat despite not having hands to put it on with!

FIRST FACIAL FEATURES

MINI STATS

Theme
LEGO® NINJAGO®

Year
2016

First appearance
The Lighthouse Siege
(70594)

Rarity

This former pirate once sailed the Ninjago seas, before he was turned into a Djinn – a magical, wish-granting being. Nadakhan's armoured torso bears the minifigure skull and crossbones symbol of the Sky Pirates, Nadakhan's new pirate crew. He caused trouble in two 2016 LEGO sets before vanishing!

FLYING PIRATE!

NADAKHAN

AT LEAST I DON'T HAVE TO BUY SHOES.

1 Ponytail hair plume piece was new in 2016

2 Belt printed on torso as his lower body doesn't have hips!

3 One of his four hands is a hook

SCURVY SEA SNAKE

The Serpentine are Ninjago bad guys, and so are Nadakhan's Sky Pirates – Clancee is a combination of the two. He has a snake head and a peg leg! He is part of the Sky Pirate crew in The Lighthouse Siege set.

TWO TORSOES

Like fellow foe to the Ninja Lord Garmadon, Nadakhan has a torso extender that gives him two extra arms.

179

This LEGO NEXO KNIGHTS lava monster is always heavy-handed in battle – because he has enormous stone fists! Moltor by name, molten by nature, his black rock body is oozing with red-hot lava, from his creepy cracked smile down to his toe claws!

BOULDER FISTS!

FIRST FISTS

Moltor's "boulder" fists were first seen as gorilla fists on G'Loona in the LEGO® Legends of Chima™ theme.

MINI STATS

Theme
LEGO® NEXO KNIGHTS™

Year
2016

First appearance
Moltor's Lava Smasher (70313)

Rarity

MOLTOR

I'VE GOT A STRONG HANDSHAKE.

1 Printed cracks and lava continue on back of head

2 Shoulder armour has four orange spikes

3 Oversized fists fit onto regular minifigure hands

LAVA TWIN

Bad news – Moltor has a twin brother, and he's even fierier than Moltor! Flama has a fiery tail, burning armour, and a transparent flame piece over his red-hot head.

PRISON-BREAK PIRATE

PRISON-BREAK PIRATE

> MY GREATEST ENEMY IS WOODWORM.

① Bicorn pirate hat features a skull with an eyepatch!

② Unique head print with eyepatch, handlebar moustache, and gritted teeth

③ Striped prison-uniform shirt with prisoner number

④ Peg-leg element, unchanged since 1989!

CAPTAIN SOTO

This peg-legged pirate prisoner was once the captain of the *Destiny's Bounty*, and later resurrected by villainous Lord Garmadon! He appears in just one 2016 set, when he breaks out of a high-security Ninjago prison.

INMATE ID

This Giant Stone Warrior is another inmate in the Kryptarian Prison, where the worst Ninjago criminals are held. His striped shirt has the same prisoner number as Soto!

MINI STATS

Theme
LEGO® NINJAGO®

Year
2016

First appearance
Kryptarian Prison Breakout (70591)

Rarity

181

1. Bucket helmet first seen on LEGO knights in 2010

2. Chain-mail printing on torso and legs

3. Shield features a red-eyed, angry bear – yikes!

MY MACE IS ALSO HANDY FOR CRACKING NUTS!

MEAN AND GREEN

DRAGON DISCIPLE?
Led by a Dragon Wizard, the Dragon Knights appeared in sets from 2010.

FRIGHTENING KNIGHT

The Frightening Knight is wearing spiky new shoulder armour and holding an even spikier new mace weapon. His bucket helmet with its distinctive dark-green plume hints that he may be a relic from the lost LEGO® Castle Dragon Kingdom.

MINI STATS

Theme
LEGO® Minifigures

Year
2016

First appearance
LEGO Minifigures Series 15

Rarity

FUTURISTIC KNIGHT
Aaron Fox from LEGO NEXO KNIGHTS is equally armed and dangerous in his Battle Suit (set 70364) – a frighteningly futuristic fighting machine with a bow and shield.

CHAPTER TEN
SPECIAL SKILLS

AS IF BEING A LEGO® MINIFIGURE ISN'T COOL ENOUGH, SOME HAVE NICHE SKILLS THAT MAKE THEM EVEN MORE AWESOME!

In polished silver and gold, Lance looks every inch the charming knight in shining armour – and he loves it! As the most vain member of the LEGO NEXO KNIGHTS team, Lance has a double-sided head that has a winking face on the other side – perhaps for when he looks in the mirror!

CELEBRITY KNIGHT!

MINI STATS

Theme
LEGO® NEXO KNIGHTS™

Year
2016

First appearance
Lance's Mecha Horse (70312)

Rarity

LANCE

IT'S HARD WORK LOOKING THIS GOOD.

FAMILY CREST
Lance's family symbol is a horse, so his torso is printed with a pentagonal horse crest.

1 Pointed visor fits over a motorcycle helmet piece

2 Silver shoulder armour with transparent orange pads

3 White and silver armour printing with orange circuitry

LANCEBOT

Lance has a faithful robot assistant who ensures he always looks his best! Lancebot is by his master's side in the Mecha Horse set.

ONLY ORANGE NINJA

1 Printing shows Japanese symbol for the number six

2 Other side of head is printed with an orange mask

3 Quiver of arrows – she also has a crossbow to fire them with

SKYLOR

NAME THE FLAME

Skylor's outfit colour is known as "flame yellowish orange" on the LEGO colour palette.

BAD DAD

Skylor's father is the villainous Master Chen. His elaborate headdress made from a snake skull marks him out as the sneaky leader of the Anachondrai tribes.

Until this colourful newcomer entered the world of LEGO NINJAGO in 2015, it was protected by just five brave Ninja. In her flame-coloured orange robes, Skylor is the skilful Elemental Master of Amber. She can absorb the powers of others – including her fellow Ninja – making this trailblazing female Ninja a real force to be reckoned with!

MINI STATS

Theme
LEGO® NINJAGO®

Year
2015

First appearance
Condrai Copter Attack (70746)

Rarity

185

I HAVE BRAIN FREEZE BUT I HAVEN'T EATEN ICE CREAM!

FROZEN POWERS

1 Transparent energy layer and hood are one piece

2 Distinctive beauty mark visible through face hood

3 Wave of water motif on shoulder harness

4 Printed azure-blue sash tied around waist

NYA (FS)

SECRET SERIES

"FS" stands for Secrets of the Forbidden Spinjitzu, which is Series 11 of the LEGO NINJAGO TV series.

When the Ninja of Water, Nya, travels to the frozen Never-Realm, her Water elemental powers turn to ice! Now her Ninja hood is topped with transparent-blue frozen water energy. This version of Nya first appeared in a 2019 set.

MINI STATS

Theme
LEGO® NINJAGO®

Year
2019

First appearance
Katana 4x4 (70675)

Rarity

SAMURAI X

Before Nya was a Ninja, she adopted the alias Samurai X so she could join the Ninja team on their missions. Now there's a new Samurai X – Nya's android friend, P.I.X.A.L.

MINI STATS

Theme
LEGO® NEXO
KNIGHTS™

Year
2016

First appearance
Ultimate Macy
(70331)

Rarity

This brave knight of the LEGO NEXO KNIGHTS realm has supercharged her battle skills to become Ultimate Macy. This variant of the royal princess-turned-fighter is exclusive to one 2016 set. Her armour has turned a powerful shade of red, with new transparent details.

ULTIMATE UPGRADE

ULTIMATE MACY

I'M WELL RED.

1 Pointed visor covers her face in battle

2 Shoulder armour in transparent red for the first time

3 Dragon crest of the royal family of Knighton

ROYAL DAD

King Halbert, Macy's father, doesn't like Macy being a knight. He hates danger, but he does enter battle in his golden armour in The King's Mech (set 70357).

187

In robes emblazoned with glistening moons and stars, the Wizard is every inch the magical sorcerer! He may share his long wizard whiskers with Gandalf the Grey from LEGO® Lord of the Rings™, but his enchanting look is mostly one-of-a-kind!

MAGIC NEW HAT!

MINI STATS

Theme
LEGO® Minifigures

Year
2014

First appearance
LEGO Minifigures
Series 12

Rarity

ROBE RARITY
The Wizard was the first male in the Minifigures series to have a slope piece instead of regular legs.

AND FOR MY NEXT TRICK...

WIZARD

① New conical hat with metallic printing

② Three-piece staff includes telescope and jewel elements

③ Two-piece fabric cape has moons and stars on reverse

④ Beard and moustache piece fits around neck

MAGICAL MATE
Another spellbinding character materialized in Minifigures Series 14. Wacky Witch has a new crooked hat-and-hair piece to match her toothless, crooked grin!

POWERFUL ONI MASK

ULTRA VIOLET

1 Oni Mask of Hatred fits over her black biker helmet

2 Purple eyes and lips

3 Black leather biker trousers with belt chains

MINI STATS

Theme
LEGO® NINJAGO®

Year
2018

First appearance
Ninja Nightcrawler (70641)

Rarity

When biker-gang member Ultra Violet dons her Mask of Hatred, she becomes invincible! Her rocky shoulder armour can repel the weapons of her Ninja foes. It's a good thing she only appears in one LEGO set.

MEAN MASKS

There are three Oni masks in LEGO NINJAGO sets, each with a different power: the purple Mask of Hatred, the red Mask of Vengeance, and the orange Mask of Deception.

I NEVER TYRE OF CYCLING.

PEDAL POWER

1 New bicycle helmet with windswept hair attached

2 Tyre-tread print across lycra cycling jersey

3 Cycling shorts, knee pads, and green boots on legs

BONUS BIKE
The Mountain Biker's accessory is, of course, a mountain bike – in a new blue and gold colour scheme.

MOUNTAIN BIKER

Rough terrain is the name of the game for the Mountain Biker! She handles off-road obstacles with ease on her heavy-tread bike. She is dressed to pedal quickly and comfortably in her tight lycra outfit, and the other side of her head reveals sporty sunglasses for when she's ready to hit the trail.

ROCKING IT
This Rock Climber from the People Pack – Outdoor Adventures (set 60202) is another lover of the great outdoors. He has a similar helmet in white, without the hair!

MINI **STATS**

Theme
LEGO® Minifigures

Year
2019

First appearance
LEGO Minifigures Series 19

Rarity

MINI STATS

Theme
LEGO® NINJAGO®

Year
2014

First appearance
Nindroid MechDragon
(70725)

Rarity

The first four Master Wu minifigures all had the same white beard and a wise expression. But that all changed in 2013, when the character got a tech overhaul and turned against his Ninja recruits! This darker variant of Wu, with a black beard and silver hat, is found in just one NINJAGO set.

HATS OFF!
Wu was the first minifigure to wear a conical hat.

MASTER OOH!

EVIL WU

DESTROY... ALL... NINJA!

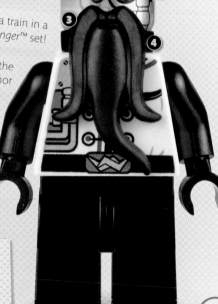

1 Wu's silver hat is part of a train in a LEGO® *The Lone Ranger*™ set!

2 Wu's cyborg implants are the work of sinister snake Pythor

3 Most Wu variants have a short white beard printed beneath the moulded beard – even this one!

4 Beard element fits between minifigure head and torso

WU'S A PRETTY BOY
Happily, Wu became good again in 2015, and turned up with his usual white beard and gold hat in the DK book LEGO NINJAGO *Secret World of the Ninja*.

191

The Ice Emperor rules over the wintry Never-Realm of the LEGO NINJAGO theme, with leadership skills that are as chilling as his armour. His white samurai helmet and body armour have built-in shards of solid ice. He has also has a secret that will make you freeze: he is really the Ninja of Ice, Zane!

ICE ARMOUR!

MINI STATS

Theme
LEGO® NINJAGO®
Year
2013
First appearance
Castle of the Forsaken Emperor (70678)

Rarity

ICE EMPEROR

DO NOT TURN ON THE RADIATOR.

1 Solid-ice crest with icicle horns

2 Transparent-blue ice head with frosty glare

3 Armour also seen on his henchman, General Vex, but in dark grey

NICER ICE

With her snowy white hair, icicle crown, and sparkly cape, the Ice Queen from Minifigures Series 16 also embodies the beauty of winter. Just watch out for her icy blasts!

KNIGHT IN TRAINING

I'M NOT YOUNG, I'M JUST VERTICALLY CHALLENGED!

ROBIN

1 Knight's bucket helmet with eye slit

2 Freckly, boyish face under helmet

3 Crest is two crossed wrenches

4 Short legs, as he is younger than the other knights

CHICKEN TO GO

The first variant of Robin has a chicken as his family crest! His new crest represents his talent for all things mechanical.

MINI STATS

Theme
LEGO® NEXO KNIGHTS™

Year
2016

First appearance
The Black Knight Mech (70326)

Rarity

Robin is a promising student at the Knights Academy who longs to be a fully fledged NEXO KNIGHT. He may be fresh-faced and short-legged, but this variant of Robin can be found in battle inside a towering Mech – as the brave Black Knight.

AVA PRENTIS

Ava is Robin's classmate at the Knights Academy. Dressed in more relaxed lavender clothing, she has short legs like Robin. She first appeared in Merlok's Library (set 70324).

I FEEL PRETTY ZEN!

COLE (ZX)

1 Silver crest on cowl represents Cole's ZX status

2 Shoulder armour includes slots for two swords

3 Lightweight armour printing

4 Back of torso shows golden Earth Dragon symbol

SHOULDER TO SHOULDER

Cole's pauldrons are also seen on the scientist Baxter Stockman from LEGO® Teenage Mutant Ninja Turtles™.

Cole ZX is the third variant of the Ninja of Earth. He is back in black, but this time he is "shouldering" heavy-duty armour. These daunting-looking pauldrons have a function as well as style – they can hold a pair of katanas across Cole's back. Looks like he means business!

KENDO ATTITUDE

Cole's Kendo minifigure sports a different kind of body armour. It has a protective plate covering his front and back torso, as well as a mask with a white grille. En garde!

MINI STATS

Theme
LEGO® City

Year
2016

First appearance
Airport Air Show
(60103)

Rarity

This wing-walking daredevil is one of a pair of stunt pilots soaring through the skies at the Airport Air Show. He wears an old-fashioned leather aviator helmet with flying goggles on his head, and a flame-emblazoned flight suit that matches his teammate.

TEAMWORK
The Stunt Pilot flies a biplane with a female stunt pilot. While one of them pilots, the other wing walks!

STUNT PILOT

YEAH, I CAN LOOP THE LOOP.

1 Aviator cap and goggles have been worn by minifigures since 1998

2 Slightly cocky grin – and rightly so!

3 LEGO City "Aviation Airshow" logo

4 Flames continue on the back of his flight jacket

① ② ③ ④

JET SET

Part of another high-flying double act in the Airport Airshow set, the Jet Pilot rockets through the air in a red jet. Her flight suit bears an image of her plane on a vertical ascent.

The **2013 version** of this NINJAGO super-baddie took his intimidating look to new heights! As well as an extra torso piece (and arms!) with ridged armour, Lord Garmadon is equipped with the imposing Helmet of Shadows, made up of three LEGO pieces.

FOUR-ARMED FOE!

MINI STATS

Theme
LEGO® NINJAGO®

Year
2013

First appearance
Temple of Light (70505)

Rarity

LORD GARMADON

I SHALL DESTROY THE NINJA ONCE AND FOR ALL!

WELL ARMED
The LEGO® Teenage Mutant Ninja Turtles™ Robo Foot Ninja also has this torso extender.

1 Clip-on horns also used on staff

2 Bat-like wings are part of a chin guard that clips onto a samurai helmet

3 Torso extender first worn by Lord Garmadon in two 2012 sets

4 Rib cage and purple sash printed on torso

BACK TO HIS SENSEI

The Master Garmadon minifigure from 2014 shows the man beneath the helmet, when he makes a brief move away from villainy!

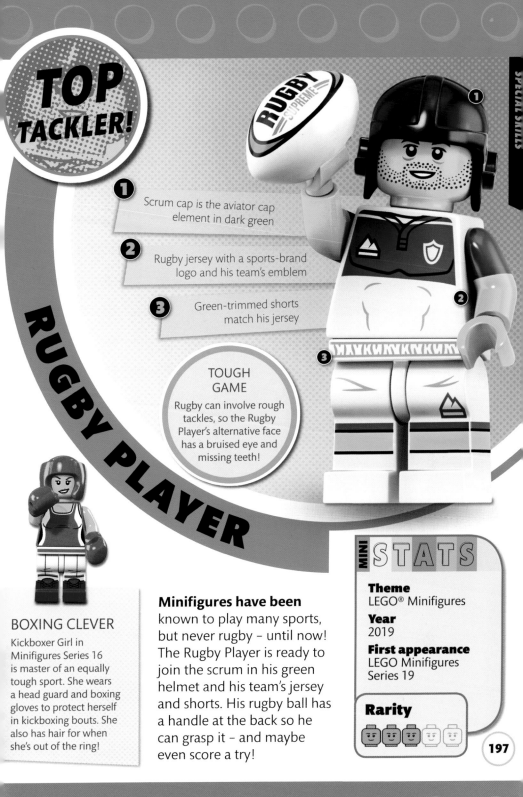

TOP TACKLER!

RUGBY PLAYER

1 Scrum cap is the aviator cap element in dark green

2 Rugby jersey with a sports-brand logo and his team's emblem

3 Green-trimmed shorts match his jersey

TOUGH GAME

Rugby can involve rough tackles, so the Rugby Player's alternative face has a bruised eye and missing teeth!

BOXING CLEVER

Kickboxer Girl in Minifigures Series 16 is master of an equally tough sport. She wears a head guard and boxing gloves to protect herself in kickboxing bouts. She also has hair for when she's out of the ring!

Minifigures have been known to play many sports, but never rugby – until now! The Rugby Player is ready to join the scrum in his green helmet and his team's jersey and shorts. His rugby ball has a handle at the back so he can grasp it – and maybe even score a try!

MINI STATS

Theme
LEGO® Minifigures

Year
2019

First appearance
LEGO Minifigures Series 19

Rarity

197

TIME WAITS FOR NO ONE... EXCEPT ME!

1 Copper clock and turning gears on hooded cowl

2 Cape shared by bounty hunter Boba Fett in LEGO® *Star Wars*™ sets

3 Red right arm and black left arm – his twin has the opposite

4 Red, silver, and copper armour with timekeeping devices

KRUX

What time is it? That doesn't matter to Krux, because he has the power to control it! He is one half of a pair of Elemental Masters of Time, also known as the Time Twins. When wearing his cowl, he is distinguishable from his twin brother, Acronix, by the old-fashioned hourglass on his torso.

198

MINI STATS

Theme
LEGO® NINJAGO®

Year
2017

First appearance
Dawn of Iron Doom (70626)

Rarity

YOUTHFUL TWIN

The Time Twins are the same age, but Acronix has a younger-looking face under his cowl because he spent years stranded in time. He also has a smartphone on his torso!

MINI STATS

Theme
LEGO® NINJAGO®

Year
2015

First appearance
Dojo Showdown
(70756)

Rarity

Blink, and **Griffin Turner** may be gone in a flash, because he is the Elemental Master of Speed. He has the ability to run incredibly fast – perhaps the reason for his swept-back hairstyle! In his white robe with red stripes, Griffin is dressed to compete in the Tournament of Elements. This minifigure appeared in just one set, then sped away.

SLICK SPEEDSTER

GRIFFIN TURNER

I PUT THE "SHOW" IN "SHOWDOWN".

1 Reflective red aviator sunglasses and a cocky smile

2 Wooden "bo" staff is a LEGO bar piece

3 Silver shuriken under belt is also a stopwatch

RARE METAL

Another entrant in the Tournament of Elements, Karlof is the Elemental Master of Metal. His time in the tournament was short-lived, as was his time in LEGO sets: he only appears in the Dojo Showdown set.

When Zane is wounded in a battle, his robotic inner workings are exposed for the first time! Showing why the Ninja of Ice is so cool, this ninth minifigure variant of Zane features metallic mechanical parts and appears in just one NINJAGO set.

ROBOT IN DISGUISE

Battle-damaged Zane is a variation on the Rebooted Zane minifigure, found only in Destructoid (set 70726), from 2014.

MINI STATS

Theme
LEGO® NINJAGO®

Year
2014

First appearance
NinjaCopter (70724)

Rarity

ZANE (BATTLE-DAMAGED)

HEAVY METAL? I'M MORE INTO COOL JAZZ!

ZANE AGAIN!
There are more than 40 variants of the Zane minifigure.

❶ Hair piece also worn by Agent Swift in LEGO® Ultra Agents

❷ Unique face print reveals the robotic workings inside Zane's head

❸ Torso print continues on back with Ice Dragon logo on tattered robe

❹ Mismatched "titanium" arm

AN ICY LOOK!

COOL CODER

PROGRAMMER

NEW TECH
The Programmer comes with two techie accessories – a sleek white laptop and a robotic dog!

WHAT'S YOUR WIFI CODE?

01001100
01000101
01000111
01001111

1 Braided black hair in a high bun was new for 2019

2 Smiling face print with round, green spectacles

3 Red check shirt tied around hips

The Programmer spends most of her time talking to technology! Her job is to write binary code, which is a numbered language that computer software can understand. She is clearly a LEGO fanatic in her free time – her black T-shirt spells out "LEGO" in code!

GAME ON!

Fellow tech fan Video Game Champ appeared in Minifigures Series 19, clutching a console and "Play-Box" game. He must take gaming very seriously – his headset is built into his green hair.

MINI STATS

Theme
LEGO® Minifigures

Year
2019

First appearance
LEGO Minifigures Series 19

Rarity

I'M NOT A MINIFIGURE!

Wondering why one of your favourite LEGO® characters isn't in this book? Perhaps it isn't a minifigure! Most minifigures are made from three standard parts: a head, a torso, and a pair of legs. Any LEGO character that doesn't include at least two of those parts doesn't get to call itself a minifigure. Let's meet some of them...

SKELETONS

LEGO Skeletons rattled onto the scene in 1995. Most have minifigure heads, but a bony handful – such as Samukai from LEGO® NINJAGO® – have special head moulds and no standard minifigure parts at all.

ROBOTS

Like skeletons, many robots and droids of the LEGO world have no standard minifigure parts. Some, like these LEGO NINJAGO droids, use other parts in unusual ways. For example, Auto (left) has a hosepipe-nozzle arm and binoculars for eyes!

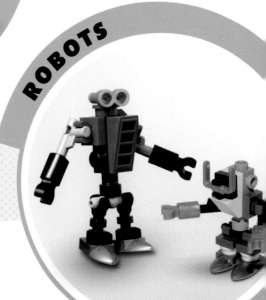

Created for the LEGO® Friends theme in 2012, Mini Dolls also feature in the LEGO® Elves, and LEGO® *Disney Princess*™ themes. Just like minifigures, they have separate heads, torsos, and legs, but in more realistic proportions.

MINI DOLLS

BIG FIGURES

Some big characters need a big figure that stands taller than the average minifigure. This Giant Troll from Fantasy Era LEGO® Castle is one example. Many "big figs" appear in licensed LEGO themes, such as The Hulk from LEGO® Marvel Super Heroes.

CREATURES

Though a LEGO monkey does have standard minifigure arms and hands, most LEGO creatures need entirely new elements to look the part. LEGO horses, cows, dogs, and even alligators all have their own special moulds.

INDEX

BY CHARACTER NAME

INDEX

DK | Penguin
Random
House

Editors Beth Davies, Hannah Dolan
Designers Lauren Adams, Guy Harvey
Production Editor Marc Staples
Senior Production Controller Lloyd Robertson
Managing Editor Paula Regan
Managing Art Editor Jo Connor
Publisher Julie Ferris
Art Director Lisa Lanzarini
Publishing Director Mark Searle

Written by Hannah Dolan
Additional text by Jen Anstruther, Jonathan Green,
Simon Guerrier, Kate Lloyd

First published in Great Britain in 2020 by
Dorling Kindersley Limited
One Embassy Gardens, 8 Viaduct Gardens,
London SW11 7BW
A Penguin Random House Company

10 9 8 7 6 5 4 3 2 1
001–320968–Sept/2020

For the curious

www.dk.com
www.LEGO.com

DK would like to thank Randi Sørensen, Paul Hansford, Martin Leighton Lindhardt, and Tara Wike at the LEGO Group.
DK would also like to thank Marc Staples for adapting files, Gary Ombler for minifigure photography,
and Lori Hand and Helen Murray for proofreading.

Content in this book has previously appeared in LEGO *I Love That Minifigure*. Thanks to Andy Jones, Rhys Thomas,
Rosie Peet, Toby Mann, Scarlett O'Hara, Tori Kosara, Ellie Bilbow, Nathan Martin, Thelma-Jane Robb,
Jade Wheaton, Amanda Ghobadi, Gary Hyde, Simon Hugo, and Markos Chouris for their work on that book.